KU-244-826

THE
SPACE
SHUTTLE
HANDBOOK

SPACE SHUTTLE

by
Kenneth Gatland
Mark Hewish
Pearce Wright

Executive editors
David Jefferis
Christopher Maynard

General editor
Kenneth Gatland

Editorial
Jacqui Bailey

Authors
Kenneth Gatland
Mark Hewish
Pearce Wright

Illustrators
Philip Green
Terry Hadler
Stephen McCurdy
Michael Roffe

Design
Stephen McCurdy
Mustapha Sidki

Photography
Michael Dyer Associates
Jerry Mason

Acknowledgements
The editors wish to thank the
following individuals and
organisations for their invaluable
help in preparing the Space
Shuttle Handbook.

Boeing Aerospace Company,
British Interplanetary Society,
Counsel Limited, ERNO
Raumfahrttechnik, Tim Furniss,
General Dynamics, Grumman
Aerospace, McDonnell Douglas,
Martin Marietta Aerospace,
NASA, Revell (G.B.) Ltd., Cathy
Cowen and Rockwell
International, Spar Aerospace
Limited, Thiokol Corporation,
Vought Corporation, OTRAG,
Spaceflight Magazine.

The Space Shuttle Handbook
was conceived and designed by
David Jefferis and Christopher
Maynard, 28-32 Shelton Street,
Covent Garden, London WC2

Produced by Sackett Publishing
Services Ltd., 2 Great
Marlborough Street, London W1
© 1979 Maynard & Jefferis
Publishing/Sackett Publishing
Services Ltd.
Published by The Hamlyn
Publishing Group Limited
London · New York ·
Sydney · Toronto
Astronaut House, Feltham,
Middlesex, England
Printed and bound by Henri
Proost, Turnhout, Belgium

All rights reserved. No part of this
publication may be reproduced
by any means without the
copyright holders' written
consent.

ISBN 0 600 38418 7

THE
SPACE SHUTTLE
HANDBOOK

Technical terms

In this book, certain expressions will crop up that may be unfamiliar to readers who are new to the world of space-flight. A brief summary follows.

Attitude The position of a craft in relation to its surroundings; it may be upside down, sideways and so on.

Escape velocity The speed (40,000 km/h) at which a vehicle will break free from the Earth's gravitational pull and head out into space.

Geostationary The high (36,000 km) orbit in which the speed of a vehicle's rotation matches that of the Earth. From the ground, the spacecraft seems to be hanging motionless in space. Geostationary orbit lies on a plane around the equator.

G force Rapid acceleration and deceleration generate what are known as 'g forces'. On Earth, the g force on a body at rest is one. The Shuttle accelerates into orbit at a rate that produces forces as high as 3 g.

Mach number The speed of sound, or Mach 1, is used as a yardstick for measuring the speed of other fast-moving objects in the atmosphere. Sound travels at 1,225 km/h at sea level. This speed decreases with altitude to a steady 1,062 km/h above 12,000 m.

Orbital velocity The speed (28,000 km/h) that will place a spacecraft into orbit around Earth.

Pitch, roll and yaw The movements of a craft that angle it away from its line of travel (see diagram p. 59).

Propellant Rockets burning liquid propellants always have two tanks; one of fuel and the other of oxidizer. The latter provides the essential oxygen without which the fuel could not burn. There is, of course, no oxygen in space.

CONTENTS

SPACE SHUTTLE

1 WHAT IS THE SPACE SHUTTLE?

The Shuttle is the first spacecraft that can be used again and again. It heralds the start of regularly scheduled flights into space.

A Swedish 'space buff' from Stockholm hopped a trans-Atlantic economy fare jet, flew 9,600 kilometres to Los Angeles, drove to Palmdale to spend two hours looking at the final assembly of the Space Shuttle Orbiter *Columbia* ... and stayed the night. Next morning he drove back to LA and caught the return flight to Sweden.

"I just had to do it," said 33-year old Ate Facrell, a computer technician with the Swedish Government. "It's the most important programme in the world. The Space Shuttle means everything to all of us. And I touched it ... it was the best day of my life!"

What is it that makes people so excited about a stubby-winged craft which takes off vertically like a rocket, man-oeuvres in orbit like a spaceship and flies home like a glider?

The answer is simple; the Shuttle is an immense leap beyond the era of one-off space extravaganzas. Flown by a crew of two or three astronauts, and capable of being reused at least 100 times, it promises to make space accessible in a regular and relatively cheap way. Its designers are convinced that such vehicles will become as vital to our economic future as ships and planes have been in the past.

The transition from the adventurous, pioneering spirit of the Apollo moonflights to the current intensely practical exploitation of space is reflected throughout the Shuttle project. One of the most succinct descriptions of this new attitude appears in an interview in July 1975 with Dr. Myron Malkin, director of the Space Shuttle programme at the Kennedy Space Center. As Dr. Malkin put it; "We have designed the Shuttle as a kind of big space truck, to give us routine access to space with large payloads. We are

▲▲ The Space Shuttle is launched like a rocket. On its way into orbit, it sheds the booster rockets and fat fuel tank. In space, its orbit can be changed by firing manoeuvre engines; they are also needed for braking to leave orbit. Gliding back through the atmosphere, the Shuttle lands like a conventional plane.

interested in making routine access possible because we see a demand for the application of space technology creeping up on us in just the same way as television and aircraft travel grew. But it has got to be much cheaper than existing systems. The name of the game is to slash launch costs for each pound of payload placed into orbit ... and the Shuttle's calculations use 1971 as a baseline, when it looked possible to cut launch costs of $1,000 a pound with existing technology, to $100 a pound with the new system."

The Shuttle System

Specialists in the aerospace business are fond of using the word 'system' and this is certainly a suitable term for the Shuttle. Two of its three main components – the Orbiter and the booster rockets – are built to standard specifications, like any assembly-line product, that make them interchangeable from one mission to the next.

The Shuttle has been described as a 'maid of all work' space freighter. The workhorse, the rocket-powered Orbiter, launches satellites, space probes and any other cargo up to a maximum weight of 29,500 kg and capable of fitting into a cargo bay 18 m long by 4.5 m wide. It can be used to rescue astronauts stranded in orbit, to retrieve and repair satellites which have failed and to launch military reconnaissance satellites from the cargo bay.

In spite of its reassuring resemblance to an aeroplane, there is nothing ordinary about the Shuttle. It is an untried spaceship being put through its paces with a human crew aboard; and it takes us out of the 'steam age' of space travel when launch vehicles crashed to destruction with each shot, and into the age of regularly-scheduled flights into orbit.

2 ANCESTORS OF THE SHUTTLE

The roots of the Shuttle can be traced back to the science fantasies and grandiose projects of the rocket pioneers of 40 years ago.

When the war in Europe ended, the Allies were astonished to find German plans for a winged rocket that could climb into space and glide back to Earth. The German rocket programme of World War Two had a dual purpose. On the one hand, the military were pressing for a new super-weapon with which to strike at the enemy; on the other, were the rocket engineers with their unquenchable enthusiasm for space flight. Yet from here, in the middle of a war, there came the ideas that were to culminate four decades later in the launching of the Space Shuttle.

The skip-glider

The story begins before the war with Dr. Eugen Sänger, one of the great pioneers of rocket flight. He embarked on a unique research programme to develop a long-range rocket bomber for Germany at his Research Institute for the Technique of Rocket Flight. Here, he and a small team set about designing and building the first engines. These test models were expected to lead to a full-size engine of 100 tonnes thrust.

Working with Dr. Irene Bredt, the brilliant mathematician whom he later married, Sänger developed the concept of a giant 28-m long rocket bomber able to fly halfway around the world. The design was quite remarkable. Its wings were wedge-shaped, flat on the underside and merged with a flat-bottomed fuselage so that the craft could glide at supersonic speeds.

It was a daring idea – and very advanced for its day. However, the project died in the summer of 1942 when it came into conflict with the ever-increasing demands of the German war machine for men and materials.

The Peenemünde rockets

In contrast to the skip-glider, the development of ballistic rockets at Peenemünde was given high priority under the direction of Major-General Dornberger and Dr. Wernher von Braun. Their first A-4 rocket, with a thrust of 25 tonnes, rose from its pad in 1942. This, of course, was the famous V-2 rocket with which the Germans bombarded southeast England. The V-2 carried a one-tonne warhead a distance of up to 305 km.

The first move to extend the rocket's range was to fit the basic A-4 airframe with swept wings and enlarged rudders. In theory, this winged rocket – called the A-4b – would travel 595 km on a skip-glide path similar to that of the Sänger-Bredt rocket. However, it never went past the initial test stage.

Target America

At the end of the war, plans were found among the mountains of documents from Peenemünde for large, two-stage rockets. A set of blueprints showed a dart-winged craft designated the A-9, mounted on the nose of a booster of more than 181,500-kg thrust. The drawings were dated 1941 and 1942.

Like the A-4b, the A-9 was meant to fly a skip-glide trajectory with a range of up to 5,000 km. With a suitably uprated booster there was no reason why it could not have attacked targets on the Atlantic coast of the United States from launch sites in western France.

Discussing the circumstances of these studies after the war, von Braun explained: "We computed that the A-4b was capable of carrying a pilot a distance of 645 km in 17 minutes. It might have taken off vertically like an A-4 then landed, glider-fashion, on a medium-sized airstrip!"

"We even designated it A-4b to give it the benefits of the A-4's high priority."

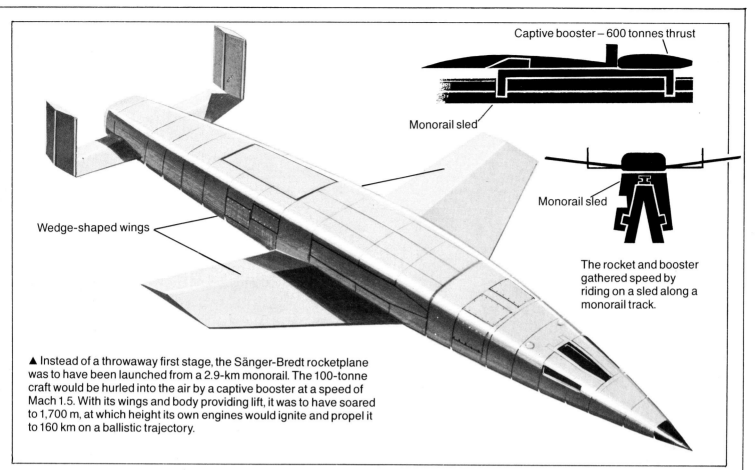

Captive booster – 600 tonnes thrust

Monorail sled

Monorail sled

The rocket and booster gathered speed by riding on a sled along a monorail track.

Wedge-shaped wings

▲ Instead of a throwaway first stage, the Sänger-Bredt rocketplane was to have been launched from a 2.9-km monorail. The 100-tonne craft would be hurled into the air by a captive booster at a speed of Mach 1.5. With its wings and body providing lift, it was to have soared to 1,700 m, at which height its own engines would ignite and propel it to 160 km on a ballistic trajectory.

◄ The two test flights of the A-4b at Peenemünde in 1944 were a limited success, though it was the first winged guided rocket ever to break the sound barrier. The first rocket failed entirely. The second made a successful climb, but spun into the sea as it returned to the atmosphere.

▼ The A-4b's re-entry into the atmosphere would have been halted by aerodynamic controls; they would cause it to bounce back into space much like a flat stone skipping across the waters of a pond. Finally, it would have glided back to Earth in a gentle path to land 750 km down-range. An uprated version, the A9-A10, would have been able to bomb the east coast of the United States from Europe.

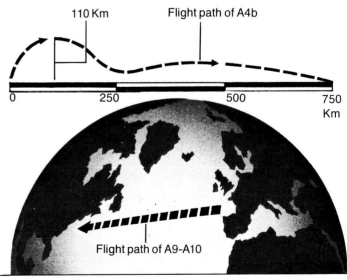

110 Km

Flight path of A4b

0   250   500   750
                    Km

Flight path of A9-A10

"If the A-9, the dart-winged variant, were to be mounted as a second stage on a powerful booster rocket, the A-10, it would become a supersonic plane capable of crossing the Atlantic. The A-10 booster was actually a very early design, and with it in view in 1936, Peenemünde's test facilities were designed to handle thrusts of up to 200 tonnes – some eight times that of the A-4."

"Beyond the A-10," von Braun said, "we had not dared to go except in imagination, although there was in our minds another still larger booster, possibly to be designated the A-11."

"Thus, with better propellants, we might easily have projected the pilot of our A-9 into a permanent orbit around the Earth!"

"His return would have been contingent upon the A-9 producing a very short power impulse to throw it into an elliptical path intercepting the Earth's atmosphere after half a circumnavigation. After a long, supersonic glide, the pilot would reduce his speed below that of sound, then he would lower his flaps and wheels and make a conventional aircraft landing."

America enters the scene

After World War Two, America devoted most of its attention to expendable rockets for both military use and space research purposes. However, a few projects, such as the rocketplanes of the X-series, did point another way to the future. These craft served as a means of gaining experience in high-speed flight within the atmosphere and gave designers a chance to tackle the problems of aerodynamic control and heating.

On October 14, 1947, Charles Yeager became the first man to fly faster than sound (Mach 1.06) in the Bell X-1, propelled by a 2,727-kg thrust RMI rocket engine. Other versions of this rocket-powered research aircraft followed – the X-1A, X-1B, X-D and X-1E. As well as conventional aeroplane steering, they also had rocket controls that enabled them to steer at high altitudes and in rarefied air.

The Bell Bomi

Major-General Dornberger – who had become a consultant to the Bell Aircraft Corporation after the war – was keen that Peenemünde's pioneering work with winged rockets be continued. He was particularly intrigued by the ideas of Sänger-Bredt and sought to take them a step further.

Under his impetus, a reusable spaceplane concept emerged in 1951 that embodied most of the principles which appeared in the NASA designs for the Space Shuttle some 20 years later.

Called the Bell Bomi, this two-stage craft had five large rocket engines in its winged booster and three in its small upper stage. They ignited as one to launch the heavy combination vertically into flight, all eight engines being fed from tanks in the booster. Two minutes and ten seconds after lift-off the upper stage separated while the piloted booster glided back to base. Meanwhile, the small upper stage, now burning propellants from internal tanks, continued on its flight.

A wealth of ideas

Interest in spaceplanes was by no means limited to America. Europe chipped in with a wealth of ideas. As early as 1948, a small study group of the British Interplanetary Society concluded that routine space operations would most likely require a delta-winged orbiter capable of returning to a runway intact, ready to be used again. At the time, however, it was assumed that its first stage booster would not fly back under its own steam but be recovered by parachute.

Other European concepts included a number of serious studies by the Royal Aircraft Establishment, Bristol Siddeley and Rolls-Royce in Britain. They examined the prospect of reusing boosters that were fitted with air-breathing engines to enable them to cruise back to base.

A variation on the theme of reusable boosters was the subject of a study by ERNO of West Germany, in conjunction with Nord Aviation and SNECMA of France. It envisaged a 52-m long, air-breathing booster, fitted with wings for take-off from an airfield. It developed 72 tonnes of thrust and carried a winged 26-m long orbiter.

The latter was propelled by six rocket engines burning liquid oxygen and liquid hydrogen. At Mach 7 and an altitude of 35 km, it would separate from the booster and continue into orbit.

Meanwhile Junkers of Germany was working with a space transporter based on the ideas of Sänger. The aim was to place a 2.75-tonne payload into space at an orbital height of 300 km. The two-stage winged rocket, weighing slightly less than 220 tonnes at take-off, would have been boosted into flight by a captive rocket sled.

MUSTARD and CRESS

Another study, partly financed by the British Ministry of Aviation, was the MUSTARD project. The name was an acronym of Multi-Unit Space Transport and Recovery Device. A small test model – CRESS – was intended for hypersonic flight-testing in the atmosphere.

The favoured launch method involved three winged rockets of similar size and shape. They were to take-off in a

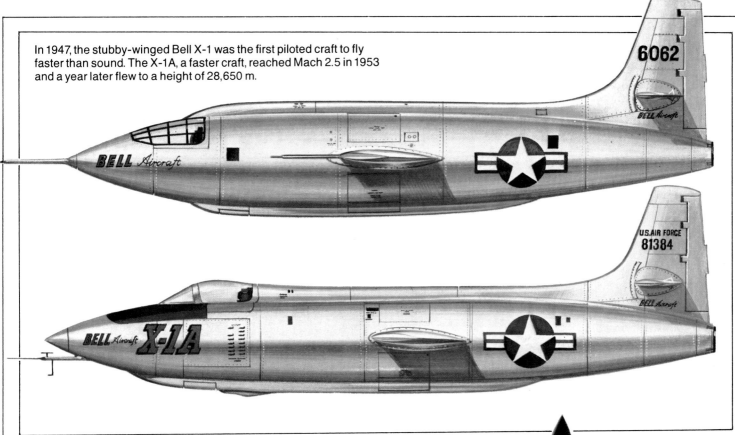

In 1947, the stubby-winged Bell X-1 was the first piloted craft to fly faster than sound. The X-1A, a faster craft, reached Mach 2.5 in 1953 and a year later flew to a height of 28,650 m.

6062

U.S.AIR FORCE
81384

'vertical stack' formation. Only the centre rocket went into orbit. The others merely served as boosters; they peeled away at 61,000 m when their task of assisting the orbiter was done. After delivering a 3,000-kg payload into orbit, the orbiter would re-enter the atmosphere and, like the two boosters, fly back to land on a runway assisted by small turbofan engines. Thus, the entire launch system would be recoverable. Of all the schemes that evolved in the 1960s, this seemed the most attractive.

The Triamese

Although none of the European projects was adopted for development, they undoubtedly played their part in shaping ideas for a practical space ferry.

Years after MUSTARD had been shelved, the British Aircraft Corporation saw the basis of their design resurrected in the Triamese. This system had three almost identical rocket-powered vehicles that were to take off in a 'stack' formation. The two boosters acted as fuel tankers and fed all the engines during launch so that the third vehicle could continue into orbit with full tanks after separation.

The day of the Dyna-Soar

Russia's triumph in launching the first artificial satellite in October 1957 had a shattering effect on the United States' military establishment. The sudden importance of the military and civilian role of space at last opened the taps for

▲ In theory, Bomi could attain speeds of up to 8,450 km/h. Bell Aircraft envisaged the small upper stage reaching a peak altitude of 44 km before beginning its glide back to Earth. Its trajectory would enable it to fly across the United States in just 75 minutes.

substantially funded projects.

In November 1959, the Boeing Company in Seattle received a contract from the Air Force to develop a manned space glider.

This delta-winged craft, known as Dyna-Soar, had both conventional aerodynamic controls for flying in the atmosphere and gas-jets for operation in space. Because of the punishing temperatures it would have to endure when re-entering the atmosphere, it was necessary to build it from entirely new high-temperature resistant super-alloys, ceramics and graphite materials.

Although the Titan III was selected as the launcher as early as February 1962, serious trouble still lay ahead. As technical difficulties grew, initial ideas of developing Dyna-Soar as a weapons system gave way to the more sober aim of simply proving the ability of a spaceplane to fly.

Dyna-Soar was now designated X-20 – and Boeing began the preparations for airdrops in 1965. For the space trials that were scheduled to follow, the vehicle would be launched unmanned on the nose of a Titan IIIC, with the aim of reaching a maximum height of 80 km. On the return leg of the journey, the vehicle would glide back to Edwards Air Force Base, its landing target in California. If all went well, the Air Force intended the first manned flight to take place in early 1966.

According to Boeing's calculations, the Dyna-Soar pilot would have been able to shorten or extend his range and to manoeuvre thousands of kilometres left or right of his flight path to reach his landing base.

The death of the Dyna-Soar

In the meantime, clouds in the shape of escalating costs had blown up on the horizon. Dyna-Soar had weight problems and the design team had become ever more aware of the enormous teething problems they faced.

It appeared that ballistic capsules could already perform many of the tasks ascribed to Dyna-Soar, including the all-important one of Earth observation. The axe finally fell on December 10, 1963 just as assembly was started. Secretary of Defence, Robert McNamara, declared that the Air Force would henceforth concentrate on a Manned Orbital Laboratory (MOL) which would employ a Gemini spacecraft as the re-entry vehicle.

MOL was in many ways a crash project born of the need to obtain the best possible surveillance of the Soviet Union and other parts of the world. Astronaut Gordon Cooper had surprised everyone in May 1963 by his vivid descriptions of Earth from orbit. When his Mercury capsule passed over North Africa, he saw what he thought was the wake of a boat

▲ MUSTARD: The motors of all three stages (above) fired together at launching. Each element was 27 m long and 18 m wide and weighed 136,000 kg at take-off. The boosters, each manned by a pilot, had a 560-km range and touched down at 190 km/h.

▼ The Triamese (below) was an American study similar in conception to MUSTARD. The orbiter separated from the two boosters prior to insertion into orbit.

The Space Shuttle Handbook

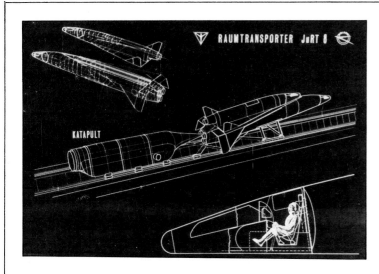

▲ The Junkers 'Raumtransporter' was to be launched by a rocket sled catapult, hurtling along an inclined track 2.9 km in length. The piggyback upper stage would have been piloted by one man.

▶ It was proposed to rocket Dyna-Soar to the edge of space on the nose of a modified Titan ICBM. The delta-winged vehicle would have attained speeds of 21,240 km/h.

on the Nile. Above India and Tibet, he spotted winding roads, houses and wisps of smoke from chimneys.

Observations from Gemini spacecraft were equally impressive. It was the aim of MOL to develop the fledgling science of space reconnaissance using astronauts to monitor the surveillance equipment.

Probing the edge of space

At the same time, the X-15 programme of the 1960s was probing the frontiers of space and exploring the same problems that the Shuttle would have to contend with. Flight trials with the first of three X-15 research planes began in 1959 with a drop from a modified B-52 bomber. The aim of the research programme was to achieve speeds in excess of six times that of sound (6,400 km/h). To maintain control in the thin upper air – where the tests would take place and where normal aerodynamic surfaces had no effect – a stick at the pilot's left hand would apply pitch and yaw control through rocket thrusters in the nose, and roll control with thrusters in the wings.

As early as November 1961, Robert M White had flown to 6,585 km/h. In July 1962, he gained his astronaut's wings by taking an X-15 to 103,269 m. In August 1963, a NASA test pilot rocketed to an altitude of 107.8 km, while in October 1967, an X-15 fitted with two external tanks clocked a speed of 7,272 km/h.

But the programme was not without its misadventures. Major J Adams, USAF, lost his life on his seventh flight, on November 15, 1967, when (it appears) he misread the roll indicator for sideslip. Re-entering the atmosphere on the

▲ The X-15 flew higher (over 100 km) and faster (to Mach 7) than any other rocket-powered plane. Several X-15 pilots earned astronaut wings by virtue of having flown to the edge of space.

▶ The rocket-powered X-24B weighed 2,268 kg. It was fitted under the wing of a B-52 parent craft prior to being carried aloft. Air-launchings conserved fuel and were used for all research flights.

wrong heading, the aircraft lost stability, stalled and broke up at 18,288 m.

The ultimate achievement of the X-15 research flights was their contribution to the understanding of hypersonic flight at the very fringes of space. The X-15 programme was concluded in November 1968.

The lifting bodies

A new generation of stubby little test vehicles, known as lifting bodies, had grown out of hypersonic wind tunnel tests at the Langley Research Center at the same time as the Dyna-Soar programme was underway. These craft were an attempt to solve a major impasse with re-entry from space. Blunt-nosed bodies, such as the Gemini and Apollo capsules with their ablative heat shields, could withstand the fierce heat of re-entry but had virtually no flying or manoeuvring capability. On the other hand, winged craft such as the X-15 became spectacularly hot as they plummeted back into the atmosphere at hypersonic speeds. The wingless lifting bodies were a happy compromise. They produced less heat during re-entry but maintained the ability to fly to a controlled touchdown.

The M2-F1 was the first lifting body. It was built of plywood around a tubular frame and was used simply to test

the new wingless shape in low-speed conditions.

On the first flight in March 1963, the M2-F1 was towed along the dry lake bed of Edwards Air Force Base behind a car. It reached flying speed, but bounced so badly that it was only airborne for a moment. Although turbulence from the tow-car was blamed, it was also clear that the craft had to be modified to attain better control. During the second test in April the M2-F1 took to the air without a hitch and reached a speed of nearly 195 km/h.

In all, the tiny M2-F1 made some 90 proving flights and set the scene for the more ambitious experiments to come.

The heavyweights

The initial success with lifting bodies led to contracts being placed with private industry in 1964 for the construction of two heavyweight test vehicles – the M2-F2 and the HL-10.

Both were launched, like the X-15, from a modified B-52 bomber flying at 13,7000 m and at a speed of 720 km/h. After being dropped, the pilot ignited the rocket engine to gain extra speed and altitude. After the engine had shut down, the pilot flew the lifting body along the same corridor that a returning spacecraft would take if it were landing at Edwards Air Force Base. During the final approach, the

The Space Shuttle Handbook

Ancestors of the Shuttle

▲ The frail little M2-F1 (above) was the first craft to show the feasibility of wingless piloted gliders. The X-24A (above right), one of the last of this breed, could glide at supersonic speeds.

▼ The HL-10, seen landing and parked, had a flat underside and rounded top, a reversal of the M2-F2 shape. It proved to be one of the most successful of all lifting bodies.

pilot increased his rate of descent. Some 300 m above the ground he pulled up the nose to perform a 'flare out' manoeuvre that halted the steep descent and cushioned the landing.

The test programme began in March 1966 with a captive flight of an engineless M2-F2 under the wing of a parent B-52. Four months later the craft unhooked for the first time at 13,700 m and glided to a 320-km/h touchdown on the dry lake bed.

In December 1966, the HL-10 was tried out but it proved difficult to control, especially in making turns. It was a case of 'back to the drawing board' and another 15 months of preparation passed before the HL-10 was ready to fly again.

In the meantime, the M2-F2 had been taken aloft for a test drop. It was recognised that the craft had to be nursed gently

through its manoeuvres because of its tendency to 'Dutch roll' – rock from side to side – in cross-winds.

In May 1967 the craft was put to its severest test when it flew into a light cross-wind. Coming out of the final turn that would line it up with the dry lake runway the M2-F2 began a violent Dutch roll. The pilot fought and managed to wrestle the craft back into control, but he was now off-course. Without the aid of runway markings he could no longer judge his altitude and the chase planes which normally provided callouts by radio were no longer in position. To make matters worse, the rescue helicopter was now too close for comfort.

Running out of height and with no time to get back on track he was forced to begin the landing flare. Hitting the lake bed just as the flare was completed the M2-F2 bounced

▲ The M2-F2, here being followed down by an F-104 chase plane, was much bigger and heavier than the M2-F1. It was dropped by a B-52 rather than towed up by a plane like the M2-F1.

▼ The X-24B accelerated to 1,600 km/h and climbed to 21,000 m before gliding to land at a brisk 320 km/h. It was the last of the rocket-propelled experimental craft that began with the Bell X-1.

25 m with the undercarriage partly extended, hit for a second time, skidding sideways, and finally rolled over tearing off the main gear, cockpit, canopy and the right fin. The pilot, though badly hurt, escaped with his life.

However, the HL-10 made better progress under rocket power. In August 1970 it attained an altitude of 23,000 m and a speed of 1,630 km/h. The HL-10 went on to achieve the fastest flight speed of any lifting body – 1,965 km/h and a record height of 27,500 m.

The last of the lifting bodies

The X-24 was to be the last of this generation of manned test vehicles. When it first appeared it had the shape of a bulbous wedge, rounded at the top, flat on the bottom and with angled stabilisers at the rear. Its first flight from a B-52

took place in April 1969.

Subsequent flights, at first unpowered, then later with a rocket engine fitted, allowed a team of Air Force and NASA pilots to test the new aerodynamic shape at speeds up to 1,687 km/h and altitudes to 21,300 m. The shape of the X-24A was later changed to that of a delta and the craft reappeared as the X-24B.

The final goal of the X-24B test programme was to try what had never been attempted before. Instead of landing on the broad dry lake bed, an attempt was made to put down on a regular runway.

So well did the X-24B respond that the pilot put the craft down only 1.5 m from the prescribed touchdown point. More flights were made to gain experience, before the X-24B landed for the last time on November 26, 1975.

3 THE SHAPE OF WINGS TO COME

"I would forecast that the next major thrust into space will be the development of an economical launch vehicle for shuttling between Earth and the installations which will be operating in space."

It was with these words, at a London meeting of the British Interplanetary Society in August 1968, that Dr. George Mueller, NASA's Associate Administrator for Manned Space Flight, chose to reveal publicly the agency's enthusiasm for building a Space Shuttle.

The vehicle Mueller described – although quite novel – was to change radically over the next four years and proved to be one of the hottest of political hot potatoes NASA has ever had to deal with.

However, in the form Mueller envisaged, it had the shape of a great flaring delta. Its deep centre-body, which was fitted with a spacious cargo bay, mounted three powerful rocket engines at the rear. Wrapped in a V-formation around its nose and wings were expendable fuel tanks that were to be jettisoned on the way into orbit. The Shuttle would take-off vertically, but would land on a runway like a conventional airliner.

The design takes shape

Late in 1968 the idea of a reusable spacecraft was officially adopted by NASA; but an enormous amount of basic research was needed before it became a viable proposition.

At the close of the decade, as NASA basked in the glory of the first moon landing, the Shuttle was just one of a host of ambitious projects – and a rather minor one at that. Competing for attention was a 12-man space station, a giant 100-man orbiting base, a moon station and a mission to Mars.

But the belt-tightening mood that settled on Washington in the early 1970s radically affected the space agency's optimistic plans. For the first time in its history, NASA was obliged to practice moderation.

The only projects to survive the initial round of budget slashing were the 12-man space station and the Shuttle. Of the two, the Shuttle clearly played a supporting role as a low-cost supply system for the space station. At the time the Saturn 1B, the rocket that supplied Skylab in 1973-74, was burning up $120 million with every launch. It would clearly be an intolerable financial burden for a space station that required a regular ferry service.

Design studies

In 1969, NASA awarded design study contracts for Phase A to the aerospace industry for a two-stage, fully reusable system that comprised a large manned booster and a smaller manned orbiter. The major constraint was the cost of boosting astronauts and their equipment into space. At the time it amounted to about '$1,000 a pound'. NASA was looking to slash this figure to 'between $10 and $50 within ten years'.

The proposals that emerged were varied, and highly inventive. They included everything from orbiters that were lifting bodies, to those that had delta wings and even straight-winged vehicles that looked rather like aeroplanes. Some were mated belly-to-belly with their boosters, others rode on top in piggyback fashion.

Unfortunately, these initial studies coincided with an economic report that pointed out a potential disaster. The fully reusable designs lacked flexibility. If technical difficulties resulted in either the orbiter or the big booster being heavier than initially planned, the whole system would have to be taken back to the drawing board; there was no halfway

▶ Here, a one-third scale model is undergoing tests in a wind tunnel. The Orbiter's design, however, owes as much to financial considerations as to technical ones, for the Shuttle is the product of considerable political compromise.

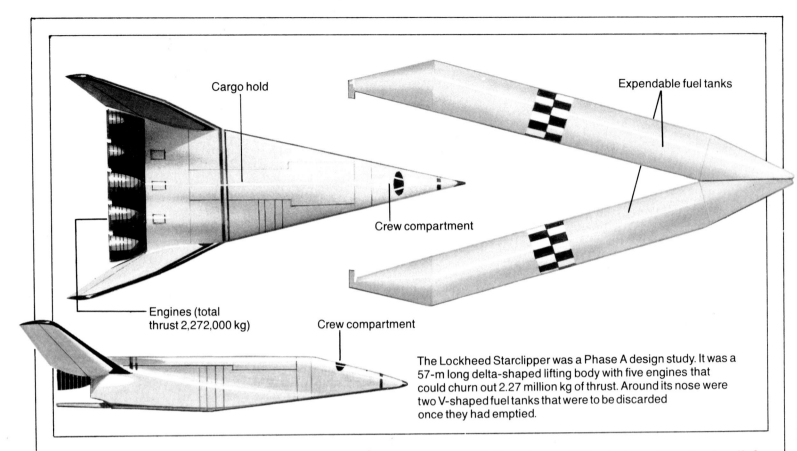

Cargo hold

Expendable fuel tanks

Crew compartment

Engines (total thrust 2,272,000 kg)

Crew compartment

The Lockheed Starclipper was a Phase A design study. It was a 57-m long delta-shaped lifting body with five engines that could churn out 2.27 million kg of thrust. Around its nose were two V-shaped fuel tanks that were to be discarded once they had emptied.

solution of tinkering with individual parts, the whole lot would have to be redesigned.

By this time it was rapidly becoming evident that linking the Shuttle with the space station project was bringing NASA within a whisker of losing both. Congress simply refused to foot the enormous bills. In 1970, a move in the Senate, in which Senator Mondale described the Shuttle as "a project in search of a mission", nearly stopped the whole programme dead in its tracks.

Choosing the delta wing

NASA dramatically changed tack and began to campaign for the Shuttle as a viable project in its own right. As justification for its existence, it was argued that it would bring huge savings to any space project, manned or unmanned. To support its position, NASA found an ally in the Air Force which was prepared to use the Shuttle's payload capability to launch big reconnaissance satellites and thus obtain a more flexible position vis-à-vis 'spaceborne threats' from the Soviet Union. But political support had a price. The Air Force insisted that the orbiter be a delta wing craft so that it would be able to make extensive manoeuvres on the way back to Earth. Thus, if a Shuttle had to land in a hurry it would not have to wait until its orbit brought it to the correct re-entry position.

Early in 1970 the choice of designs had narrowed and contracts worth some $20 million were placed with McDonnell Douglas and North American Rockwell for Phase B studies. These would define the final shape of the Shuttle and anticipate Phase C (nuts and bolts design) and Phase D (production and operation).

The favoured system that emerged was a booster the size of a Jumbo jet, carrying on its back an orbiter the size of a medium-range airliner. The total weight of the two craft was an estimated 1,587,000 kg; a figure that included the fuel tanks embodied in the airframe of each.

The Phase B Shuttle would function as follows: at launch the booster's 12 engines would lift the vehicle straight up from its pad and into its departure trajectory. At a height of 61,000 m the two vehicles – already climbing at an angle to the horizon – would separate and the orbiter continue into space under its own power. The booster would circle back under power from its turbofan engines and make a conventional touchdown.

By all standards, the Phase B design was a formidable project. The recognition that its research and development bill could climb to $10,000-$14,000 million (at 1970 prices) forced NASA, already coping with a slashed budget, to whittle down the concept. It seemed that the only way to save money, and retain a useful space programme, was to prune drastically. Three Apollo lunar landers were axed, the 12-man space station disappeared, and as for the Shuttle, it was a case of back to the drawing board yet again.

It had also become evident that the budget for the Phase B

The Space Shuttle Handbook

▲ This early Shuttle proposal by Grumman consisted of a swept-wing orbiter mounted on the back of a large flyback booster. After the two craft had separated, the booster would have cruised back to base under the power of a set of jet engines.

◄ It was Max Faget of NASA's Langley Research Center who proposed a reusable, twin-stage Shuttle design as early as 1969. He envisaged a straight-winged orbiter riding piggyback on a manned booster. Only the booster engines fired during launching.

▼ The most promising Phase B Shuttle design proved to be one of a delta-winged orbiter atop a Jumbo-sized flyback booster. Both were manned by a crew of two. The orbiter had a cargo bay large enough to have held a space station module, for at this time NASA still planned to build a 12-man orbiting station.

The Shape of Wings to Come

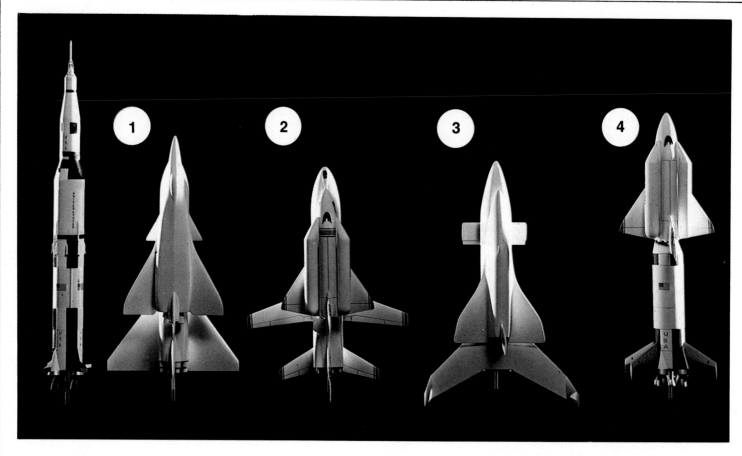

▲ Some of the various reusable orbiter and booster combinations that were considered during Phase B studies are shown here. The most likely Phase B design (1) had delta wings for both the orbiter and flyback booster. The Grumman/Boeing design (2) featured an orbiter with external fuel tanks. A swept-wing booster (3) contrasts with the delta-winged orbiter, a design insisted upon by the Air Force. The orbiter mounted in tandem on a Saturn 1C (4) was a cheap alternative to expensive flyback boosters.

Shuttle would never have made it past the White House. By mid-1971 NASA had calculated that it would need up to $2,000 million annually to fund the project. The White House was prepared to allow only half this figure.

It was a last-ditch joint study by the designers at Boeing and Grumman that appeared to provide the way out of the impasse. They adapted the big Saturn 1C, the first stage of the moon rocket, as a compromise booster for the Shuttle. By adding massive tailfins, delta wings, a nose section and flight deck, landing gear and jet engines they proposed transforming it into a reusable booster. The cost of the entire system was pared down to just $1,200 million, only $200 million more than the White House had initially indicated. But it was not good enough.

Reluctantly, NASA was forced to abandon the proposal. Late in 1971, all hope of a manned booster, and a fully reusable Shuttle system, died.

The final shape

In the meantime, on January 5, 1972, the go-ahead for the Shuttle programme had been given official blessing by President Nixon. The vehicle that eventually emerged was quite unlike the big composite of the Phase B study, however. Instead of a Jumbo-sized flyback booster, it was proposed that the orbiter should fly into space on a big, belly-mounted expendable fuel tank. The orbiter could also be reduced in size since most of its fuel was now carried externally; a solution that permitted the big cargo bay and payload capability to remain.

There was considerable debate as to whether the orbiter-fuel tank combination would be launched with liquid or solid fuel boosters. Both kinds were examined in the expectation that they would be recoverable, even though they were unmanned. In March 1972, the solid-fuel boosters were chosen. It was planned to have them splash down in the sea after use and be recovered by tug. However, if a solid fuel booster sank the loss would be a fraction of that of a liquid fuel booster. Again the decision was dictated by finances.

The spaceplane had also undergone a number of other changes. The first casualties were the two jet engines that would have assisted approach and landing manoeuvres. The decision to omit them came from a growing confidence in the orbiter's ability to glide home without power. Shuttle pilots use a highly automated landing system to put down

▲ A scale model Shuttle, complete with Orbiter, external tank and boosters, is tested in a wind tunnel to study its behaviour in flight.

◀ Here, a quarter-scale model of the Orbiter is being hoisted onto a comparably scaled fuel tank for tests that simulate the vibrations of lift-off and the early stages of flight. The dark strips are sensors that record the model's behaviour.

▼ A scale version of the Jumbo carrier and the Orbiter undergo a separation sequence in a wind tunnel, exactly as they did during the actual approach and landing tests.

▲ The first Orbiter assembled at Rockwell was one of five. However, this vehicle was simply a prototype used for tests in the atmosphere. It was to be christened *Enterprise*.

safely. There is, of course, no question of the pilots aborting a landing and going round again.

Also discarded were two 175,450-kg thrust solid rockets, originally to have been mounted on the rear fuselage. This cut the gross weight by about 3,175 kg. These back-up motors would only have ignited in the event of a failure of the orbiter's main engines. They would have boosted the Shuttle to flight speed and allowed it to dump the big external tank over the sea and glide back to base. Now reliance is placed on the solid rocket boosters to perform this function, a much safer procedure.

Though pared down from the original concept, the final Shuttle system, consisting of an Orbiter, fuel tank and boosters, still represented an enormous technical challenge. However, the changes in size and scope had reduced development costs by almost half, to $5,500 million, with an extra $1,000 million as a contingency fund in the event of unforeseen problems arising.

Originally, the Shuttle's designers had hoped to provide docking facilities in the front of the craft so that two Orbiters could link up nose to nose. But two factors frustrated this design. First there was the problem of protecting the Orbiter's nose from over-heating during re-entry; second was the large amount of space that would be taken up by the docking tunnel which would have extended beneath the flight deck.

The solution was to transfer the docking facilities to the front section of the cargo bay immediately behind the Orbiter's main cabin.

The Orbiter's engines

The Shuttle's main engines are highly advanced rocket motors. Three of them are clustered in the tail of the Orbiter. They burn liquid oxygen and liquid hydrogen supplied from the external tank. By operating at the very high pressure of 211 kg/cm^2 they generate immense power despite being relatively small.

Unlike earlier rocket engines that had a firing life of

▲ In September 1976, just four years and nine months after the official go-ahead, Orbiter-101 was towed out before the public accompanied by a band playing the theme music from 'Star Trek'.

minutes, those of the Shuttle are designed to work for up to $7\frac{1}{2}$ hours during a lifetime of some 55 missions. They have a two-stage combustion cycle. First the fuel is partially burned at low temperatures and high pressure, then it is completely burned in the main chamber. This method achieves a spectacularly high degree of combustion efficiency, over 99 per cent.

Housed in blisters alongside the main engines are two rocket motors used to manoeuvre the vessel in space. Each can deliver 2,727 kg of thrust and burns fuel stored in tanks on board the Orbiter. These motors have three functions. They nudge the spaceplane into orbit after the big fuel tank has separated, they provide manoeuvring ability in space and, when fired at the end of a mission, they serve as brakes to start the descent back to Earth.

Finally, the Orbiter is equipped with sets of tiny thrusters in the nose and next to the orbital manoeuvring engines at the rear. They are used to maintain attitude control – to swivel and position the orbiter with a high degree of accuracy. Like the manoeuvre engines these thrusters run on self-igniting (hypergolic) fuels – liquids that ignite spontaneously when mixed to produce blasts of hot gases.

How to christen a spaceplane

By the autumn of 1975 the first full-scale vehicle, Orbiter-101, was taking its final shape in Rockwell's plant at Palmdale in the Mojave Desert of California.

As the Orbiter neared completion in 1976, the question of what it should be baptized cropped up. NASA turned instinctively to the tradition of naming space vehicles after patriotic sentiments and suggested *Constitution.* However, some backstage arm-twisting that involved the White House and a concerted campaign by fans of 'Star Trek', the popular television space series, to snow President Ford with 100,000 letters produced the desired result. The spacecraft was dubbed *Enterprise,* in honour of the spaceship which features in the 'Star Trek' programme.

4 THE 'ENTERPRISE'

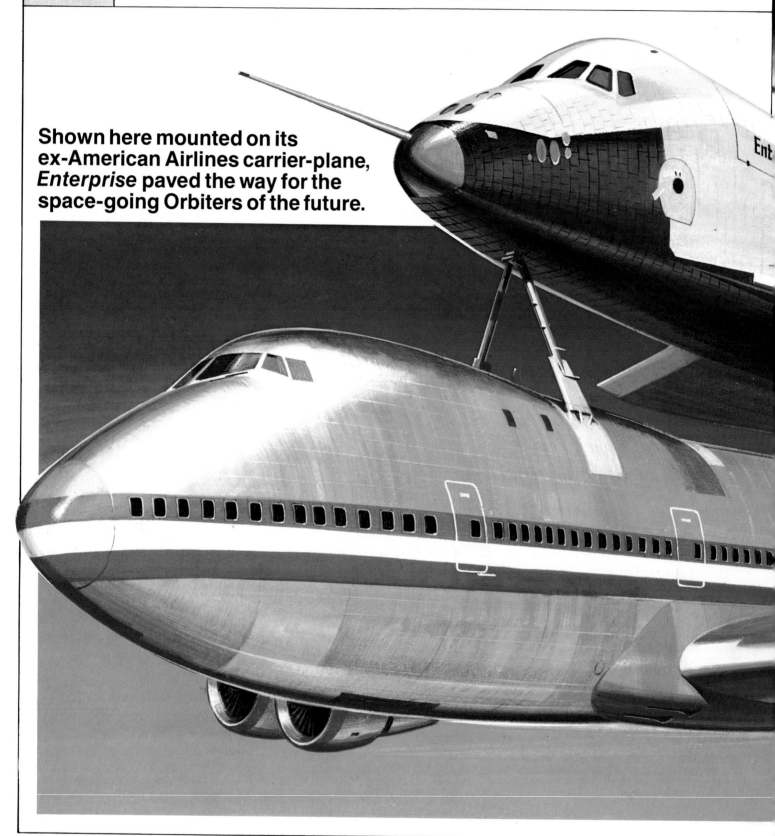

Shown here mounted on its ex-American Airlines carrier-plane, *Enterprise* paved the way for the space-going Orbiters of the future.

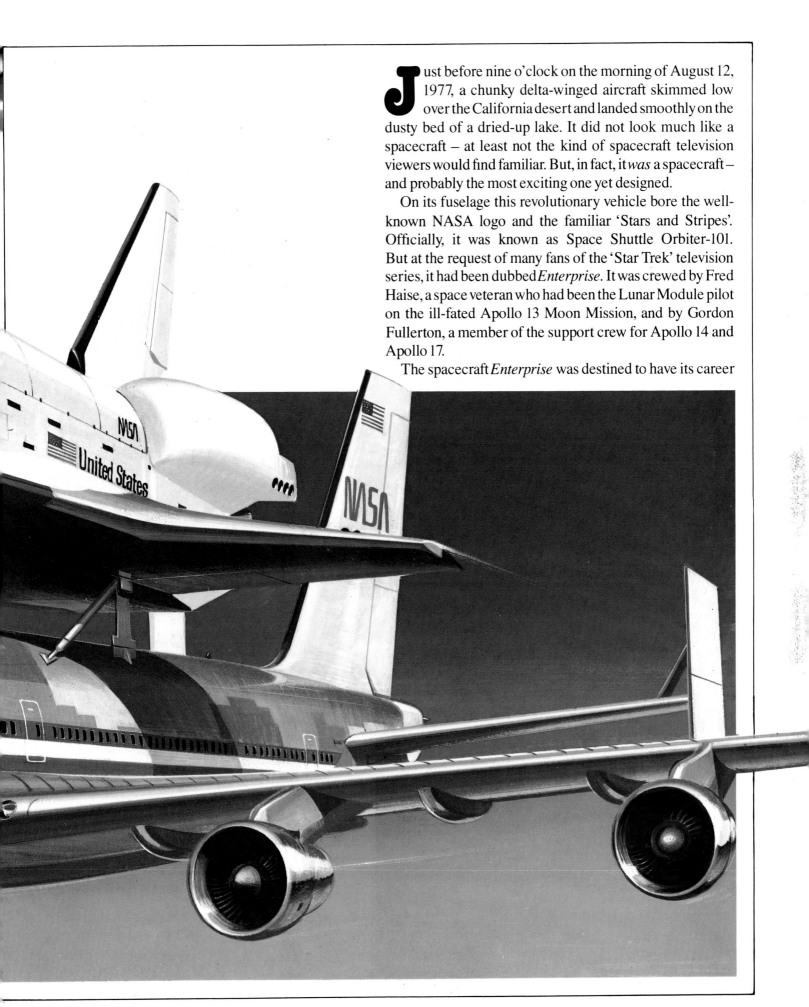

J ust before nine o'clock on the morning of August 12, 1977, a chunky delta-winged aircraft skimmed low over the California desert and landed smoothly on the dusty bed of a dried-up lake. It did not look much like a spacecraft – at least not the kind of spacecraft television viewers would find familiar. But, in fact, it *was* a spacecraft – and probably the most exciting one yet designed.

On its fuselage this revolutionary vehicle bore the well-known NASA logo and the familiar 'Stars and Stripes'. Officially, it was known as Space Shuttle Orbiter-101. But at the request of many fans of the 'Star Trek' television series, it had been dubbed *Enterprise*. It was crewed by Fred Haise, a space veteran who had been the Lunar Module pilot on the ill-fated Apollo 13 Moon Mission, and by Gordon Fullerton, a member of the support crew for Apollo 14 and Apollo 17.

The spacecraft *Enterprise* was destined to have its career

The *Enterprise*

▲ Piggyback aircraft are nothing new. The idea dates from 40 years ago when, in July 1938, the Short-Mayo Composite made the first Atlantic flight with a load of freight and mail. After both craft reached the right height and speed, the seaplane detached itself from the flying boat and flew from Ireland to Montreal.

cut short by one of Earth's most familiar problems, a shortage of money. But before it was axed, it was to prove that a reusable spaceplane was a practical proposition.

Whereas earlier American astronauts and Soviet cosmonauts had to plop unceremoniously into the sea or thud to the dirt in Central Asia, shuttlenauts will be able to step sedately from their vehicle after landing on a conventional runway.

Although *Enterprise* was known as Orbiter-101, and by implication was the first of the Orbiter family, it is now unlikely ever to join the ranks of those that leave the Earth's atmosphere. Plans to reuse it have been abandoned.

The Approach and Landing Test (ALT) programme, of which the August 12 flight formed a vital part, was one of the two major contributions that *Enterprise* made to the Space Shuttle project. After completing the second of these (the Mated Vertical Ground Vibration Test) at Marshall Space Flight Center in Alabama, *Enterprise* had fulfilled its purpose and faced an uncertain future.

Assembling the *Enterprise*

Rockwell International's Space Division began assembling the crew module for *Enterprise* at Downey, California in June 1974. In August of that year work was begun on the rear fuselage. By 1975, components from all over the United States began converging on the United States Air Force's Plant 42 at Palmdale, California. Here, *Enterprise* and all

future Orbiters are to be assembled. In March, the centre fuselage section arrived from General Dynamics in San Diego. Grumman-built wings were delivered from New York two months later and the vertical tail, supplied by Fairchild, was shipped from Long Island a couple of days after. On September 17, 1976, *Enterprise* was rolled out into the California sunshine for its first public viewing.

Approach and landing tests

Because of the nature of a Shuttle mission, NASA had to start its testing at the end of a flight sequence and work backwards. Since Orbiters are to be launched vertically from a pad in much the same way as a rocket, but land on a runway just like a plane, it was sensible to prove that *Enterprise* could be handled safely in the atmosphere at an early stage of the trials.

For test purposes, the vehicle was not fitted with any engines of its own, so some method of getting it airborne had to be found. One early proposal involved mating two fuselages of the Lockheed C-5A Galaxy, a military cargo aircraft, side-by-side. They were to be joined with a central wing section under which the Orbiter would be slung. When this

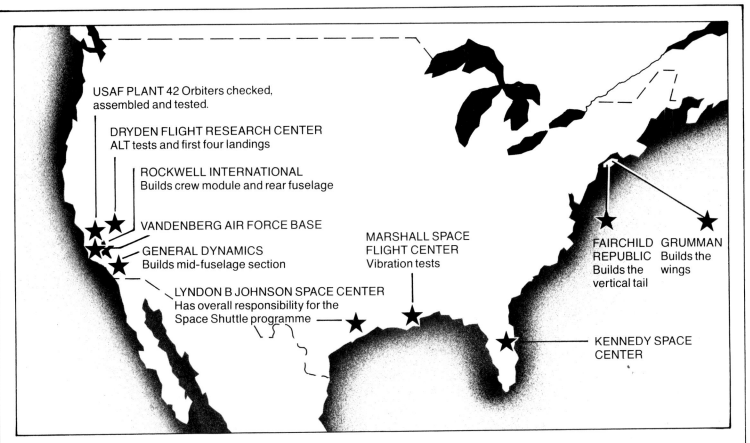

USAF PLANT 42 Orbiters checked, assembled and tested.

DRYDEN FLIGHT RESEARCH CENTER
ALT tests and first four landings

ROCKWELL INTERNATIONAL
Builds crew module and rear fuselage

VANDENBERG AIR FORCE BASE

GENERAL DYNAMICS
Builds mid-fuselage section

LYNDON B JOHNSON SPACE CENTER
Has overall responsibility for the
Space Shuttle programme

MARSHALL SPACE
FLIGHT CENTER
Vibration tests

FAIRCHILD
REPUBLIC
Builds the
vertical tail

GRUMMAN
Builds the
wings

KENNEDY SPACE
CENTER

▲ This map shows where sub-contractors for the main components of *Enterprise* are located, together with the test facilities and launch sites of the Shuttle System.

▼ The mighty Kenworth diesel truck-and-trailer combination which hauled *Enterprise* on its ground journeys was over 40 metres long and rolled on 90 wheels.

The *Enterprise*

idea was abandoned, as a result of its cost and complexity, NASA turned to the idea of using a single large aircraft. The space vehicle would be attached, piggyback fashion, to the top of its fuselage. The Galaxy was again suggested for this role but rejected. The plane chosen in the end was an ex-American Airlines Boeing 747-100 airliner.

Although the *Enterprise*-747 combination looked unwieldy, its combined weight during the tests was less than that of a fully fuelled Boeing 747 en route between London and Los Angeles.

Changes to the carrier craft

The NASA 747, named the Shuttle Carrier Aircraft (SCA), needed very few modifications for its new role. The most obvious were the mountings for the spacecraft and the installation of end-plates on the tailplane. This improved its stability with *Enterprise* attached.

The NASA 747 was delivered to the Dryden Flight Research Center at Edwards Air Force Base in California on January 14, 1977. It was joined 17 days later by *Enterprise,* which had been towed the 58 kilometres from Palmdale on a truck. A huge gantry, known as the mate-demate facility, had been built at Dryden to attach the two craft. The spaceplane was first raised by the gantry. Its landing gear was retracted and the 747 towed into position beneath it. *Enterprise* was lowered gently onto the two mounting struts on top of the airliner's rear fuselage. The right-hand one was fixed, leaving the left-hand unit adjustable for minor corrections. Then it was attached firmly and the single forward strut also secured.

On February 15, the two craft underwent three taxi tests at speeds of 253 km/h. These were to assess ground handling and control and, most importantly, to see if the 747 could steer and brake effectively with its single, oversize passenger; *Enterprise* is roughly the size of a DC-9 passenger jet!

Airborne

Three days later the combination took to the air for the first time. The Orbiter was unmanned and fitted with a specially built tail-cone over the rear fuselage to reduce drag. The 747's crew were delighted. They reported that they couldn't even tell *Enterprise* was aboard. The two craft, they claimed, handled much more like an unmodified airliner than they had expected.

A further four flights with the unmanned Orbiter completed the tests, the sixth mission proving unnecessary. During these, the spacecraft-747 combination was put through its paces to ensure that no minor fault would turn

▲ January 31, 1977: a slow-moving convoy escorts the 68,000 kg *Enterprise* from USAF Plant 42 to Dryden Flight Research Center, a trip of nearly 60 kilometres.

into a disaster. The airliner was flown with its two starboard engines reduced to idling. An emergency descent was made to within a few metres of the desert floor with all four engines idling, and two tests of the separation systems were carried out. The 747 was also braked to a halt in less than 2,000 metres of runway, thus proving it could land on short airstrips with *Enterprise*.

The next three trials were flown with the Orbiter crew in the cockpit, monitoring the performance of the vehicle. But before they took place, *Enterprise's* auxiliary power units and fuel cells, which provide hydraulic and electrical power, were tested on the ground. The Orbiter was also prepared for both emergency separation from the 747 and an emergency escape by the crew. This involved fitting explosive bolts in the attachment struts, installing pilot ejection seats and fixing a charge to blow out the overhead escape-hatch panel.

The Space Shuttle Handbook

Having completed its part of the ALT programme, the 747 will be used to ferry Orbiters between their various assembly, test and launch sites. Originally, NASA proposed fitting a power pack of four Pratt & Whitney TF33 turbofan engines to Orbiters so that they could fly across the country under their own power. The idea was abandoned, again on the grounds of cost and complexity – the two bugbears that everyone on the Shuttle project constantly battles with.

On June 18, 1977, Fred Haise and Gordon Fullerton took their places for a first taste of what it would be like to fly a Space Shuttle. The mission lasted less than an hour and was uneventful. Tests were limited to simple checks of the Orbiter's systems.

The back-up crew fly

Ten days later, the back-up *Enterprise* crew, Joe Engle and Dick Truly, took the controls for the second captive flight. After flying a simulated separation manoeuvre, the 747 dipped into a 6° glide (the normal approach angle for an airliner is 3°) so that the Orbiter's automatic landing system could be checked. Minor changes were made to *Enterprise* after these two flights. Then the number one crew took the controls for a final dress-rehearsal captive flight on July 26.

The first free flight

August 12 was the big day for which Fred Haise and Gordon Fullerton had been waiting since being selected for the ALT programme. During this time they had been honing their skills, first in NASA's Northrop T-38 Talon training aircraft fitted with a special speed brake, then in a modified twin-jet Grumman Gulfstream. Later, they had practised in the Space Shuttle Procedures Simulator and 'flown' the Orbiter Aeroflight Simulator.

That morning, the *Enterprise*-747 combination climbed away from the runway and to a height of 7,346 metres. Then, while flying at 500 km/h, the 747's engines were reduced to idling speed. Its spoilers were deflected upwards in order to increase drag. Explosive bolts on the three strut mounts were fired and the Orbiter, which had been canted 6° nose-up for this flight, separated for the very first time.

▲ Poised like a giant bat above its carrier-plane, *Enterprise* awaits its first flight. Between them, the two craft weighed more than 250 tonnes.

▶ The top picture shows the landing after *Enterprise's* first captive flight on February 18, 1977. Below, nearly six months later, *Enterprise* separates from the 747.

Three seconds later, Haise and Fullerton rolled their spacecraft 20° to the right, while the 747 turned away to the left to avoid a collision. *Enterprise* was flying free.

Soon after separation a minor hiccup occurred. One of the four main control computers in the spacecraft was voted out of operation by its three counterparts, as it was disagreeing with them. But this caused no concern and the commander described his 5½ minutes of free flight as "super slick". The vehicle came to a halt after a roll of 3,350 metres along the dry lake bed runway of Edwards Air Force Base.

The second free flight, a month later, included manoeuvres at both high and low speeds, as well as a roll to the left. This subjected the craft to an acceleration of 1.8 g, to provide information for use in the Terminal Area Energy Management (TAEM) part of Orbiter's descent.

Computer in control

TAEM is a computer programme. It constantly evaluates the Orbiter's position, attitude and velocity so that the spacecraft is guided automatically, at the desired speed and descent angle, from a height of about 21,300 metres to just over 3,000 metres. Here, the final approach to touchdown begins.

On the third free flight, the performance trials continued. While still at 1,980 metres, the crew engaged *Enterprise's* automatic landing system and on-board computers flew the spacecraft for 45 seconds. During this time the Orbiter was guided by signals from a microwave landing system on the ground.

Crew in control

It is important to keep in mind that the Orbiters are not powered during re-entry. Though stable, and able to manoeuvre well under the right conditions, their approach and landing has to be controlled automatically. Their descent is very swift, and no second chance is available without power. As one Rockwell official said, "If the crew were to jump over the side of their craft without the benefit of parachutes, the Orbiter would still beat them to the ground."

Since the microwave landing system is so vital to the success of any Shuttle mission it is not surprising that it was tested early on. The equipment is also being installed at the Kennedy Space Center in Florida and at Vandenberg Air Force Base in California, where the runways for returning Orbiters are located. Each runway measures 100 metres wide and 5,000 metres long. The microwave beam coverage

The Space Shuttle Handbook

extends from the ground up to an angle of 30° and fans out 15° from either side of the runway centre-line. The system also includes distance-measuring equipment accurate to within 30 metres. On operational missions, the Orbiter will be under the control of the microwave beam while it is between 3,000 and 60 metres above the ground.

Although the Kennedy Space Center and Vandenberg Air Force Base will be the only launching and landing sites for operational Shuttles, the first four flights will end with landings at Edwards Air Force Base in California. This has the advantage of almost limitless runways on a dried-out lake bed. It will remain a diversion field for regular Shuttle missions.

On the third free flight the *Enterprise* crew switched off the automatic landing equipment after its 45-second trial and returned to a control-stick method of steering after the vehicle had descended to 1,000 metres. With this method, the spacecraft's controls are trimmed under the command of a computer, which acts in response to movements of the stick and rudder pedals by the crew.

During the first three free flights, a fairing was fitted over *Enterprise's* tail-end to reduce drag and possible vibration. But operational Orbiters cannot use such a device, and so for the last two tests in the ALT series the fairing was removed. Dummy engines were installed to give *Enterprise* the aerodynamic qualities of an Orbiter returning from space. During the first of these flights, *Enterprise* dipped to a maximum descent angle of 28°–twice that reached with the tail-cone on. This is also steeper than would be attained on a real mission, when the angles will vary between 21° and 24.° The crew reported that, apart from increased drag, the testcraft handled virtually as it did with the fairing installed.

Finally, to complete the ALT tests, *Enterprise* was ballasted to move its centre of gravity aft to a position similar to that of orbital flights to come.

Rough landing

During this flight, *Enterprise* was piloted manually by Gordon Fullerton. Instead of the perfect computer-controlled landings of other occasions, *Enterprise* touched down on only one set of wheels. It bounced, nose high, then as Haise lowered the elevons to get the nose down, crabbed up into the air. It bounced and tipped to the right a total of three times before the shuttlenauts managed to correct the situation.

Watching the landing was a distinguished party of visitors

The *Enterprise*

including Prince Charles of Britain. Commenting afterwards on the rough landing, Fullerton remarked dryly that "...the excitement at touchdown displayed that the gliding performance was better than the engineers had predicted."

Having shown that it could be flown just like a conventional aircraft, albeit one not designed solely for flight within the atmosphere, *Enterprise* then had to demonstrate that ferry flights between bases were just a formality. The tail-cone was refitted and the Orbiter was placed back on the 747. The Boeing was fuelled as if for a typical ferry flight. In November 1977, the combination made four flights on consecutive days, the longest lasting four and a quarter hours. Following these trials *Enterprise* was moved to a hangar at Edwards Air Force Base. There it was modified in preparation for the Mated Ground Vertical Vibration Test on the other side of the continent at Marshall Space Flight Center in Alabama.

▲ *Enterprise*, and future Orbiters, touch down at about 320 km/h, slightly faster than a commercial jet. The 'nose-up' angle of landing helps the delta wings to cushion the ride.

▶ After initial vibration tests, *Enterprise* was lifted out of the Dynamic Test Stand at Huntsville. This picture shows it suspended in the air before being winched back in again, this time to be attached to the boosters and the external tank for more tests.

Vibration testing

Although *Enterprise* closely resembles production vehicles from the outside, it is nowhere near operational standard. All its engines are dummies and the fuel tanks are filled with gas under pressure rather than with the liquid oxygen and hydrogen they would normally contain. Nor are the 34,000 thermal protection tiles, developed to absorb and dissipate the heat built up during launch and re-entry, fitted. Their place is taken by plastic plates. Also, the wing leading edges of *Enterprise* are clad with glass-fibre.

The Space Shuttle Handbook

The men who flew the *Enterprise*

Fred Haise, above left, was the Lunar Module pilot for the Apollo 13 trip to the Moon in 1970. He was lucky to get back alive as the mission nearly ended in disaster. His crew-mate aboard *Enterprise* was

Charles Fullerton. Dick Truly and Joe Engle, shown below, were the second *Enterprise* crew to fly the 68,000 kilo craft during the highly successful nine-month ALT programme.

▲ This picture shows *Enterprise* being attached to the back of its carrier-plane. The two craft are surrounded by the lattice of scaffolding at the specially-built mate-demate facility at Dryden Flight Research Center.

Before the vibration tests could begin, the non-standard equipment fitted for the ALT series had to be removed. The nose boom, ejection seats and other items were taken out and replaced by a new nose cone, electronics and ballast. After being readied, the Orbiter was again fitted to the 747 and, on March 10, 1978, was ferried to Ellington Air Force Base in Texas. Here, staff from the Johnson Space Center in Houston – which has overall responsibility for the Shuttle project – were allowed to inspect their handiwork. Three days later the Orbiter continued its journey to Huntsville, Alabama. There it was mated with the External Tank (ET) and the two Solid Rocket Boosters (SRB's) for vibration testing. These trials were completed at the end of 1978 and *Enterprise* finished its role in the Shuttle programme as the pathfinder that never made it into space.

The second Orbiter, *Columbia*, has been allocated the honour of making the Space Shuttle's first flight. But NASA originally planned to use *Enterprise* later. It was once due to be the second Orbiter into space, making its maiden flight in 1981. NASA decided to adapt another of the Orbiters in its place and *Enterprise* was relegated to fifth place for use in the mid-1980s. By this sleight of hand, about $100 million was hoped to be saved; the amount needed in 1978 to overcome unforseen difficulties in the programme.

Meanwhile, *Enterprise* has been condemned by budget cuts to remain earthbound unless NASA can find the money to ready it as the fifth Orbiter.

The very fact that the Shuttle has been described as 'the DC-3 of the Space Age' – a tribute to its role as a workhorse – demonstrates its dramatic advance on single mission, throwaway launchers. *Enterprise* and its crews can take much of the credit for getting the test trials off to a good start. It would be a fitting acknowledgement of this achievement if Orbiter-101 were eventually to make the trip into space.

Naming the Orbiters

Entering into the spirit of exploration, NASA named the first four Orbiters after famous ships that have charted the oceans of the world. Orbiter-102, the first to be launched into space, is called *Columbia* after the sloop which explored the mouth of the Columbia River in 1792. Orbiter-103, *Discovery,* is named after the vessel in which Henry Hudson hunted for the north-west passage between the Atlantic and the Pacific Oceans in 1610. Orbiter-104 is *Atlantis* and Orbiter-099 is *Challenger,* both being names of ships that logged millions of kilometres conducting research and exploring the oceans.

5 THE SHUTTLENAUTS

Unlike early generations of astronauts, the men and women who fly in the Shuttle work in a comfortable 'shirt-sleeve' setting.

Although it will be many years before the average starry-eyed holidaymaker checks in at a spaceport, the Shuttle represents a big step toward making such a dream possible. But for the time-being, the crews who man this spacecraft will have a crowded schedule of far more sober activities on their minds. Marvelling at Man's conquest of a new environment is still low on their list of priorities.

The Orbiter, in which the shuttlenauts live and work, is designed to carry seven people into space and to house them for periods of up to a month, although early missions will rarely extend beyond a week. The crew of the vehicle consists of a commander, pilot and mission specialist. Up to four payload specialists may fly with them, their task being to look after the experiments on a Spacelab mission.

The Orbiter's cabin is designed as a combined working and living area. It is a pressurised compartment with a volume of 71.5 cubic metres, and it has three levels; an upper section or flight deck, a mid-section that serves as the living quarters and a bottom level packed with environmental control equipment.

The flight deck

The flight deck can seat four people during launch and re-entry. At the front sit the commander and the pilot. The commander acts as the captain of the ship. He has overall responsibility for the mission and makes the major decisions about safety and the allocation of resources to ensure the success of the flight. To his right sits the pilot. He has virtually an identical set of controls before him so in the event of an emergency either man can fly the ship and return

▲ ▶ The roomy flight deck seats four people. In the cockpit bay sit the commander (nearest) and pilot. Behind them are the mission specialist and the payload specialist (nearest). The cutaway view (right) shows the commander and pilot working at the rear 'on-orbit station'. The mission specialist works at the right-hand wall, the payload specialist on the left. Slipper-like footholds in the floor help the crew anchor themselves to keep from drifting away from their stations.

it to Earth single-handedly. This part of the flight deck, with its banks of instruments and forward-facing windows, closely resembles the cockpit of a large airliner. Seats for two other crew members, the mission specialist and a payload specialist, are located immediately behind the pilot and the commander.

Along the rear wall of the flight deck are two crew stations that face toward the cargo bay. They are used for handling payloads and controlling docking manoeuvres. The right-hand station contains the displays and switches needed to release and capture payloads. The person working here can open and close the cargo bay doors, deploy the giant cooling radiators fitted to them, operate the robot arm with which payloads are handled, and control the lights and closed-circuit television cameras mounted within the bay. The

The Shuttlenauts

operator can also look through either of two windows in the aft wall to watch what is going on.

The second crew station is for rendezvous and docking in space when, for example, satellites are being recovered. This station has controls and displays for the rendezvous radar and hand controls for juggling the Orbiter into exactly the right position in space.

To the left-hand side of the flight deck, when facing forward, is the payload station. It has a two-metre square bank of controls and displays that enables the operator to deal with the payload.

The mission station, on the opposite side of the flight deck, is used to manage the systems that link the Orbiter to its payloads. It performs a function that is vital to the safety of the vehicle. The mission station operator can communicate with both attached and free-flying payloads, and manage the Orbiter's routine 'housekeeping' functions from here.

The living quarters

Alongside both the mission and the payload stations are hatches which allow the crew to descend into the living quarters. Here, there are three seats for payload specialists to use during launch and re-entry, with room for a further three seats to be installed in the event of a rescue mission. The mid-section houses sleeping bunks, a galley and a 'bathroom' with a washing arrangement and a zero gravity lavatory. It also contains the air-lock that gives access to the cargo bay, and has a hatch that acts as the main door of the Orbiter when it is on the ground. Almost all of the 4.2 cubic metres of storage space for loose equipment is in the passenger compartment. It is used for items needed for a specific mission but which are not permanently mounted in the Orbiter cabin. They can be fitted into standard containers and loaded via the side hatch.

Haute cuisine

One of the major differences between the Shuttle and its predecessors is the standard of in-flight food. In the past, astronauts have had some rather uncomplimentary remarks to make about space menus which included paste-like food squeezed from tubes in a lukewarm condition, dehydrated foodstuffs compressed into cubes and rehydratable items, with cold mashed potatoes being voted the all-time low.

All this has changed in the Shuttle, however, which has a galley area containing a battery of kitchen equipment including a warming oven and freezer. The oven, which operates by forced air convection, can maintain a temperature of 63°C and reach a maximum of 85°C. The problem

encountered in previous spacecraft, of foods boiling even when only slightly warmed no longer exists since the Orbiter's cabin is maintained at sea-level pressure.

Some 30 minutes to an hour before each meal, one of the crew members will begin preparing the food. This consists of removing the appropriate packages from their storage lockers, injecting water (a by-product of the craft's fuel cells) into dehydrated foods, placing items that need heating in the oven and assembling the remaining portions on individual meal trays. Kitchen duty is hardly onerous, the entire process of preparing a meal takes one person about five minutes for a crew of four.

When it is ready, the hot food is removed from the oven and placed on trays. The meals can be eaten at a table in the galley, held on the lap, or even attached to the wall by fastening hooks. Ordinary utensils and cutlery are employed and are wiped off after the meal for reuse later. Leftovers are dumped into a garbage store.

A varied menu

A typical day's food supply for a crew member, consisting of three meals and extra snacks and drinks, amounts to no more than 20 packages. Recommended menus are planned to provide 3,000 calories a day. This has been raised from the 2,800 calories of Skylab to take into account the fact that Shuttlenauts use up more energy in running their larger and more complex craft. The menu is a standardised one with enough variety in it (there are 74 kinds of food) to run six days without repetition. This is a change from the personal preference diet used in the past. If a Shuttlenaut has a particular like or dislike, he or she can make substitutions from the pantry, which includes many items available directly from the shelves of supermarkets.

A typical day's menu is as follows: breakfast – chilled orange drink, peaches, scrambled eggs, sausage, sweet roll and cocoa; lunch – cream of mushroom soup, ham and cheese sandwich, tomato salad, banana, biscuits and tea; dinner -- shrimp cocktail with sauce, steak, broccoli au gratin, strawberries or pudding, biscuits and cocoa.

Shirt-sleeves in space

Shuttlenauts will live and work in a 'shirt-sleeve' environment. Inside the Orbiter the air, temperature and pressure will be virtually the same as on Earth. The crew will not have to wear space suits except when operating outside the vehicle.

These comfortable conditions are maintained by an elaborate environmental control system which removes excess heat, provides re-oxygenated air, circulates it to keep

▲ Meals in space have come to resemble those of commercial airlines. The system above, pioneered during the Spacelab programme, has food packed in aluminium boxes, with pull-top lids, that slot into a warming tray. The food is eaten with conventional cutlery when piping hot. Meals include main courses, hot and cold drinks in collapsible containers, desserts and plenty of condiments. The plastic packs on the side contain rehydrated foods – they need only have water added to make them fit to consume.

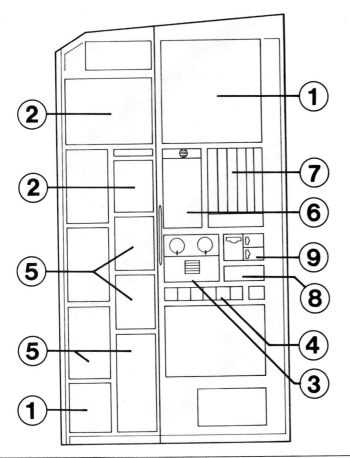

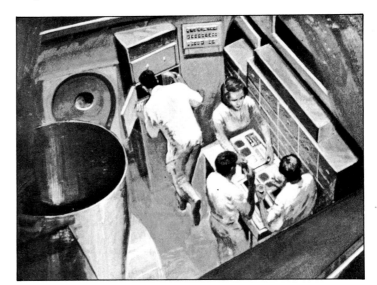

◄▲ Teams of food technologists have been feeding astronauts in space for nearly 20 years. The latest developments in in-flight catering have returned to the idea that meals should look and taste like real food. Gone are the days of compressed cubes, of paste-like goo in squeeze tubes that was eaten cold and of lukewarm rehydrated items.

Also, meals can now be consumed in a traditional manner at a table; though zero-gravity conditions make sitting down irrelevant – one can eat floating in any position.

The Orbiter galley is equipped with a freezer; its features also include: (1) Waste store; (2) Drinks; (3) Water; (4) Condiments; (5) Food store; (6) Oven; (7) Eating trays; (8) Cups; (9) Wiping towels.

it fresh and maintains a correct pressure. Orbiter, and Spacelab, are completely insulated from the cold and heat they experience in space. However, the crew, the electrical apparatus and the chemical reactions of the experiments all give off heat, and this must be removed if the vehicle is not to become alarmingly hot. Cabin air is passed through heat exchangers, where some of the heat is transferred to a water loop and the air reemerges cooled. The heat is then drawn off by another cooling loop filled with Freon (used in many domestic refrigerators) and discharged into space via large radiators mounted on the inside of the cargo bay doors. This system keeps the cabin temperature between 18°C and 27°C and, if needed, it can be augmented by the evaporation of water directly into space. The standard arrangement dissipates seven kilowatts per hour continuously – as much as is produced by seven single-bar electric fires – and can rise to a peak of 12 kilowatts every three hours.

Fresh air

The heat exchangers also control humidity. Water vapour in the air is condensed and passed to a collector. There it is sucked up by a water separator and the water/air mixture is separated in a centrifuge. The liquid is collected in a tank from where it can either be vented into space, or returned to the cabin air. This complex procedure is necessary because there is no gravity, something we take so much for granted on Earth.

Cabin air also passes through cartridges, containing carbon filters and lithium hydroxide, that remove carbon dioxide, trace gases and odours. It is then topped up with oxygen and nitrogen to maintain the normal balance and to replace losses resulting from minor leaks. There is no natural convection in a spacecraft, caused by warm air rising, because of the zero-gravity conditions. To avoid stagnation and clouds of gases forming the air is kept moving by fans.

Fire is a deadly enemy in space. The environmental control system has fire and smoke sensors which trigger off a semi-automatic fire-fighting system in an emergency. Smoke and toxic gases can be vented into space via valves if they become too concentrated, although cabin pressure is of course lost in such an event.

The environmental system can operate for 42 man-days at a time and has sufficient supplies to include three periods of outside activity by two Shuttlenauts. On a rescue mission the system might have to support up to ten people and can, in a contingency, run for an extra 16 man-days.

Shuttlenauts wearing space suits will work outside the vehicle at times to conduct certain experiments, deploy and

▼ ▲ Portholes in the rear wall of the flight deck (below) let the crew watch what is happening in the cargo bay as the robot arm unloads a satellite (above) and readies it for insertion into orbit.

The Space Shuttle Handbook

repair satellites, adjust instruments and assemble structures in space. A new design of suit has been developed by the team which was responsible for those worn by the Apollo astronauts. The suit has a stiff upper torso section made of aluminium, with a life support back-pack permanently attached, and a lower 'trousers' section. Apollo space suits were made for each individual astronaut and were flown only once. The Shuttle suits, however, come 'off-the-rack' in three different sizes that can be adjusted to fit different crew members. They are designed for a lifespan of 15 years.

Outdoor work

Shuttlenauts can don the suit, make four connections and check that it is operating correctly, all in five minutes. The space suits are stored in the Orbiter's air-lock. A crew member puts on the lower section first, then crouches under the rigid upper torso that has been hung on a bulkhead and stands up into it, pulling on the top like a sweater. The upper and lower pieces are quickly joined at the waist by means of a ring-shaped connector. This is a great advance over the Apollo design, in which the suit and back-pack were put on separately, took more than an hour to get ready, and needed two astronauts to help one another. The back-pack's seven-hour supply of oxygen, water and other expendables can be replenished in about seven minutes – less than half the time needed with Apollo suits.

Three-piece suit

The complete space suit arrangement of undergarment, pressure suit and life-support equipment is known as an extra-vehicular mobility unit (EMU). The undergarment has thin plastic tubing woven into its fabric to cool and ventilate the body. The two-section space suit has an air-tight lining covered with layers of puncture-resistant insulating fabric. A display and control unit mounted on the upper torso houses a microprocessor – a tiny computer. It provides start-up instructions, checks the suit's vital functions, alerts the wearer if something goes wrong and indicates what should be done to correct the fault – all automatically. The suit is completed by a bubble helmet, a cap with radio earphones, a microphone and a visor for eye protection, a drink bag and a waste-removal system.

The back-pack

The life-support back-pack supplies oxygen, ventilation, maintains suit pressure and cools and circulates the water in the undergarment. It also controls oxygen temperature, cleans carbon dioxide and odours from the suit atmosphere and holds an extra half-hour supply of oxygen for

▼ Crew members can slip into space suits and exit via the air-lock in the mid-section to attend to payloads. Only two crew members at a time can work outside the Orbiter.

▲ The Shuttle programme enables a broad spectrum of people to fly into space, not just young test pilots but also women, older adults and non-Americans. The latest intake of 35 trainee Shuttlenauts includes women for the first time. All six will follow the training schedule for mission specialists. Shuttlenaut Resnik (above) poses in the outfit used for walks in space. It is not custom-made, as were previous space suits, but comes in small, medium, or large sizes that fit everyone – male or female.

emergency use. A two-way radio is also contained in the back-pack.

All the joints in the suit are made of flexible fabric; this allows Shuttlenauts to move with much less exertion than was needed had they been using an Apollo suit, which had joints made of moulded neoprene rubber and cables. The gloves are made of Kevlar, a tough but thin fabric which enables space crews to handle small tools easily and to pick up items as thin as a coin.

The crews

Shuttle commanders, pilots and mission specialists will all be NASA astronauts. The latest batch of 35 fledgling Shuttlenauts, made up of 21 military officers and 14 civilians, and including six women, was selected in 1978 from a list of more than 8,000 applicants. They joined NASA's pool of 27 astronauts, a group of both pilots and scientists. Of the new candidates, 20 will become members of Shuttle flight crews. The other 15, including the women, will be mission specialists.

Initial training, which is the same for all candidates, lasts two years. It includes academic work, with studies in astronomy, celestial mechanics, life sciences, spacecraft design and general familiarisation with the Shuttle system. There is also flight training on T-38 jet aircraft.

The new Shuttlenauts complete their basic training by July 1980 at which time they are eligible for crew selection. After being assigned to a particular mission the crew spends a further six to eighteen months of intensive training before the flight.

Although the flight crews and mission specialists are all Americans, the team of payload specialists includes Europeans as well. Five potential candidates were selected in 1978, two Americans and three Europeans. The 'Euronauts' must be fluent in written and spoken English, have a

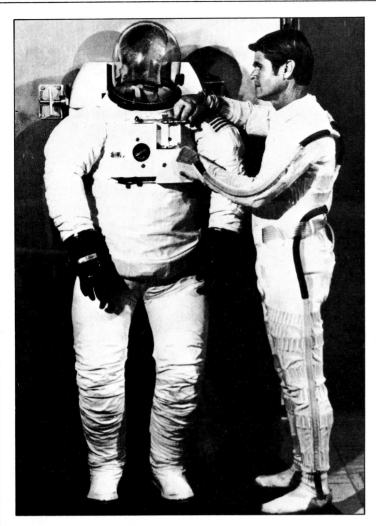

▲ An engineer models the garment worn under the pressure suit. It has a network of water-filled tubes sewn into the fabric that cool the body. The dark tubing is a venting system that conducts cooled air to the head, hands and feet.

university degree and at least five years' experience in one of the fields relevant to the particular mission. They must also have what has officially been described as 'high motivation, flexibility, emotional stability, low aggression and empathy for their fellow workers' – a demanding combination.

Payload specialists, normally appointed about a year before the flight, start their part-time training two months prior to the flight. They then transfer to the Johnson Space Center for full-time training.

For the initial flight, two mission specialists have been selected; though only one will travel into space. His prime responsibility is to oversee the Orbiter's housekeeping functions; everything from power to communications, cabin pressure, air conditioning and the countless other systems that keep life in the craft running smoothly. The training requirements for mission specialists are not as rigorous as for pilots. It is in this role that the women in the Shuttle programme will first fly into space.

Rescue in Space

▲ If an Orbiter is disabled, a second one can be sent up to rescue the marooned crew. Those Shuttlenauts with space suits will transfer their ship-mates inside portable 'rescue balls', carrying them by hand. They will manoeuvre in space using back-packs fitted with gas jets for propulsion.

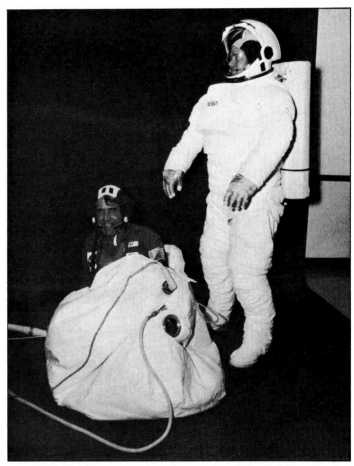

▲ Not all members on board the Shuttle have bulky space suits of their own. In the event of an emergency, a portable escape system, or 'rescue ball', can be employed. Essentially, this is a collapsible bag, 86 cm in diameter, made of three layers of tough fabric with a small viewing port woven into it. A crew member climbs inside, zips up, then turns on the life-support system to inflate it and awaits rescue. The 'rescue ball' has an intercom and a sufficient supply of air for 20 minutes.

6 A LABORATORY IN SPACE

Spacelab, Europe's major contribution to the Space Shuttle programme, is the most ambitious project yet devised to make space regularly accessible to scientists.

Once it was proved that man could survive outside the Earth's atmosphere without difficulty, the direction of America's space programme swung very swiftly from exploration to the exploitation of space. The American Skylab programme paved the way in 1973-74; although by then NASA was already looking forward to a more substantial research station; in effect, a fully-equipped laboratory in space. The laboratory would be manned by scientists, not astronauts, who would conduct research in the weightless conditions of space.

Earlier, in 1969, NASA had invited the European Space Research Organisation to share in the post-Apollo programme. Studies began the following year and led to what was then called the 'Sortie Lab'. By August 1973 the European Space Conference agreed to proceed with Spacelab, as the project had become known. The overall management of the project was placed in the hands of the European Space Agency (ESA).

Orbiting Thermos flask

Spacelab has been unflatteringly described as an orbiting Thermos flask. It is cylindrical in shape and is designed to maintain a constant temperature and pressure so that its occupants – the payload specialists who operate and monitor the experiments – can work in a comfortable 'shirt-sleeve' environment. The lab will be lifted into space in the payload bay of the Orbiter. Once there, it will operate

◄ Its magnetometer booms fully extended, Spacelab operates from its snug position in the Orbiter cargo bay. Below the glowing rear windows of the flight deck, the air-lock can be seen.

from inside the bay rather than fly free, although future versions may be left in orbit while the Shuttle itself comes back to Earth.

At work in space

NASA is planning a total of 572 Shuttle flights by 1991. Of these, 226 (nearly 40 per cent) are to include Spacelab. The laboratory consists of two basic elements; a pressurised compartment in which the payload specialists work, and a set of pallets on which are mounted the instruments that need to be directly exposed to the vacuum of space. Both elements can be assembled in a number of different combinations.

On a typical mission, the combination of pressurised compartments and pallets selected for the flight are loaded into the Orbiter's cargo bay at one of the two launch sites, either Kennedy or Vandenberg. The specialised payloads will have been installed already inside the lab. During the ride into orbit the payload specialists – up to four may travel at one time – ride in the Orbiter's cabin. Once in space, the specialists crawl through a tunnel to get to the lab. They continue to live in the main crew compartment throughout the mission, which may last from seven to 30 days, 'commuting' to work via the tunnel. Normally, only two specialists at a time occupy the lab, although at a pinch there is room for four.

A flexible shape

The pressurised compartment is just over 4 m in diameter and 2.7 m long. On most missions, two compartments will be linked together to give an overall length of 6.9 m, except when the rest of the payload bay is needed for a very long pallet load. The single module has a useable volume of slightly more than 7.6 cubic metres, the double more than 22 cubic metres. It is filled with a mixture of oxygen and nitrogen, closely resembling normal air, and kept at 21°C and at normal sea-level pressure.

The Spacelab system

Spacelab is a highly flexible structure. It can be assembled in three main ways; module only, module and pallet, pallet only. The size of the module and the number of pallets can be varied too.

The Instrument Pointing System (IPS) can aim its pallet load of up to 3,000 kg of equipment with a phenomenally high degree of accuracy – better than one second of a degree.

A twin-segment pressurised laboratory; each segment is 2.7 m long and 4 m in diameter. The double-module can carry 4,600 kg of research equipment.

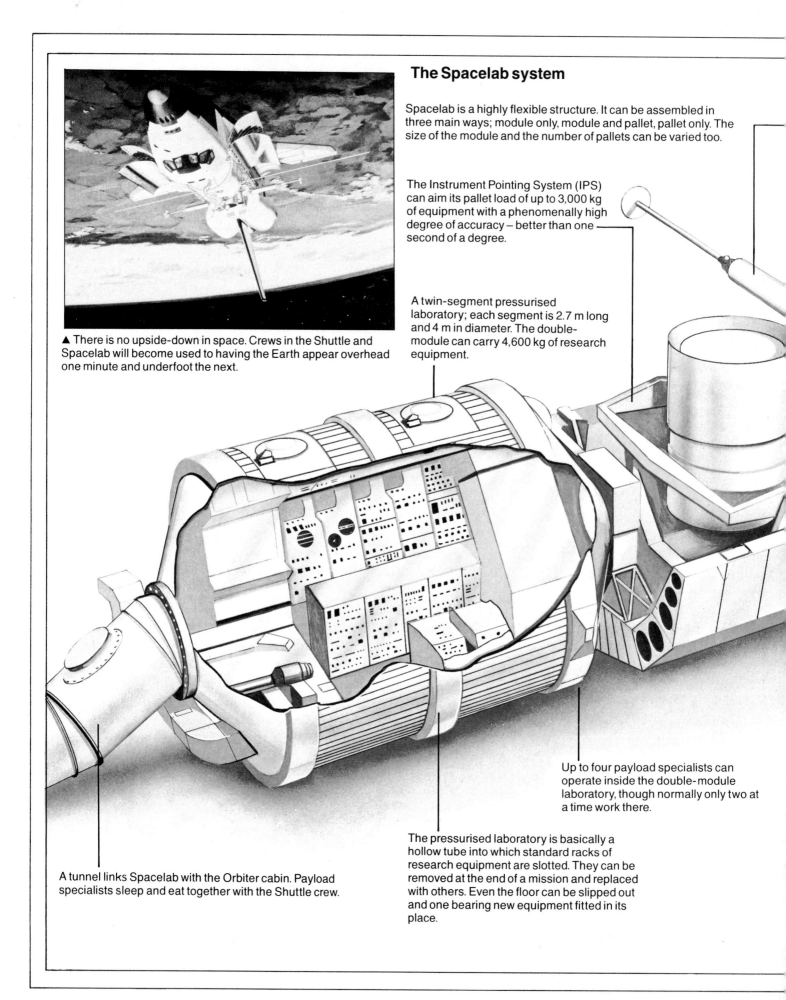

▲ There is no upside-down in space. Crews in the Shuttle and Spacelab will become used to having the Earth appear overhead one minute and underfoot the next.

Up to four payload specialists can operate inside the double-module laboratory, though normally only two at a time work there.

A tunnel links Spacelab with the Orbiter cabin. Payload specialists sleep and eat together with the Shuttle crew.

The pressurised laboratory is basically a hollow tube into which standard racks of research equipment are slotted. They can be removed at the end of a mission and replaced with others. Even the floor can be slipped out and one bearing new equipment fitted in its place.

The Space Shuttle Handbook

The sensors of the magnetometer, when fully extended on their long cable booms, will undertake geological surveys of the Earth. The data they gather can be used to locate mineral deposits of mining potential.

Lidar is a device that uses powerful laser beams to sound the little-known regions of the atmosphere between 35 and 120 km above the ground.

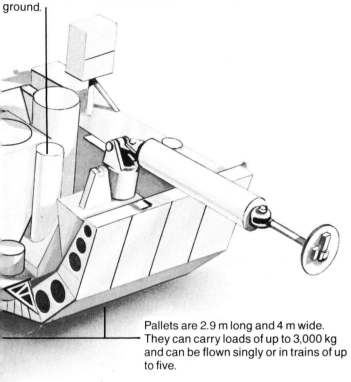

Pallets are 2.9 m long and 4 m wide. They can carry loads of up to 3,000 kg and can be flown singly or in trains of up to five.

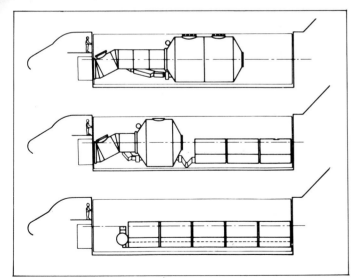

▲ Top: A twin-module laboratory with tunnel. Middle: A single-module lab with three pallets. Below: An all-pallet load with an 'Igloo' attached at one end.

Part of the pressurised section contains equipment, mounted in a series of racks, that is needed on all missions to ensure that the laboratory functions properly. This includes computers, tape-recorders and monitoring consoles. The rest of the lab will be fitted with racks of research equipment such as furnaces, microscopes, centrifuges, incubators and photographic apparatus.

Power on board

Power and other services to keep the lab comfortable are supplied by the Orbiter. An average of seven kilowatts of electricity is constantly maintained with peaks of up to 12 kilowatts if needed. All the experiments and equipment are designed to withstand accelerations of 3 g and a noise level of 136 dB (far louder than that of an airliner's jet engines) during the launch into orbit. However, when compared to the 9 g accelerations of the Apollo rocket launchers, the Shuttle by no means gives Spacelab a rough ride.

The cargo pallets are U-shaped structures some 4 m wide and 2.9 m long. Each one can carry a load weighing up to 3,000 kg and consisting of items such as telescopes, antennae, radar devices and other sensors. The maximum number of pallets that can be slotted together in the cargo bay is five, carrying a total weight of 10,340 kg; this load being the limit that the Shuttle can safely return to Earth. When the full complement of pallets is on board there is no room for a pressurised compartment. Instead, the payload specialists supervise operations from the rear flight deck of the Orbiter. Then, a special pressurised, temperature-controlled housing called an 'Igloo' is installed in the cargo bay to provide the pallet mounted equipment with electrical power and other services. Each pallet is designed for 50 flights and can withstand 3 g accelerations and temperatures that vary from --160°C to 120°C.

Paying for Spacelab

As their contribution to the joint US-European space programme, the ten member nations of ESA are providing the first Spacelab to NASA free of charge. The lion's share of the money (53.5%) is being put up by Germany with other major contributions coming from Italy (18%) and France (10%). The balance is split between the remaining members of ESA – Belgium, Denmark, Great Britain, Netherlands, Austria, Spain and Switzerland.

The first flight

Difficulties with the Shuttle's development, and with support facilities such as the Tracking Data and Relay Satellites that transmit communications between Spacelab and ground stations, threatened to delay the Spacelab timetable. By early 1979, however, ESA was pressing ahead with its goal to deliver a prototype Spacelab by mid-year. This model was used to ensure that the fully-operational lab fitted properly into the Orbiter's cargo bay and mated with all the necessary connections.

The final working model – consisting of a double segment and a single pallet – was scheduled to be delivered to NASA in two parts in the autumn of 1979, followed by an Igloo and three more pallets in January 1980.

The first Spacelab mission was scheduled originally for December 1980 but it is generally conceded that this date has slipped to the beginning of 1981. Both the first and second missions are concerned mainly with checking the laboratory as part of the overall Shuttle system.

The first orbiting Spacelab will be a modest affair comprising a single pressure module and one pallet. Because the mission is largely to test the lab itself, Spacelab 1 is only to carry 60 experiments from Europe, 15 from the USA and one from Japan; a third of its normal load. The payload specialists' work is to be limited to some 100 man-hours.

This mission will take place on the 11th Shuttle flight. It is a joint NASA-ESA venture, lasting for seven days and is to be manned by two payload specialists, one from the United States and one from Europe.

The total crew on the mission will number six men, allowing the workload to be broken down into 12-hour shifts with each team comprising a pilot, a mission specialist and a payload specialist.

Later missions

Spacelab 2, tentatively scheduled for spring 1981 aboard the 14th Shuttle flight, is an all-pallet trip sponsored by NASA. Though the primary objective is also to verify the performance of the laboratory, research in astronomy, high-energy astrophysics and solar physics is to be conducted as well. In addition, work in medicine, botany and plasma studies will help to certify the wide range of roles the lab can fill.

Spacelab 3 is presently due to be launched in 1981 too, on the 16th Shuttle flight. It is to be the first fully-operational flight and is scheduled to conduct various experiments into low gravity aspects of space. By severely restricting crew movements and the operation of the rocket thrusters, g forces in the Shuttle can be kept 1/10,000th below that of Earth for long periods.

Between 1982 and 1983, Europe plans to sponsor four Spacelab missions. Though the USA will continue to be the dominant user other countries, such as Japan, are also likely to reserve space.

Each Spacelab has a design life of 50 seven-day missions so that additional laboratories will have to be built to meet anticipated demand.

The Spacelab crew

The crew of Spacelab are known as payload specialists. They are scientists and engineers rather than highly-trained astronauts, and need only be of normal good health to make a Shuttle trip. They can be either male or female and the upper age limit has been set at 50 years. Once chosen to accompany a payload, they are trained for the flight by NASA. The training includes exercises in zero-gravity movement and detailed briefings in the flight plan, emergency procedures, in-flight operations and in-orbit housekeeping.

The payload specialists run the experiments and analyse the results using on-board computers. The presence of scientists in space is of incalculable benefit since they can modify experiments while they are in progress and even evaluate the results as they happen. If any equipment mounted outside on the pallets needs attention, or if there is a fault with the Spacelab itself, the crew can don space suits and climb into space through an air-lock to investigate the problem.

Spinoffs from Spacelab

Spacelab carries a highly flexible system of work benches and racks that can be fitted in different combinations. Here, new research methods and scientific processes will be studied, such as fusion welding and crystal growing. Here too, exotic, lightweight alloys and techniques for large-scale space manufacturing will be tried out. It may be that space processing techniques will make for such an improvement in quality that they will justify the expense of shipping the raw materials into orbit. The lab will also be used to conduct surveys of the Earth, and to observe the universe without having to cope with the barrier of the atmosphere.

Up to four Spacelab missions a year are expected to be used primarily for processing materials. The reason for this is that space, as a working environment, has three great advantages over Earth; gravity is minimal, temperature is constant and there is no atmosphere. For example, sedimentation and convection in liquids is greatly reduced in the near absence of gravity. Sedimentation (heavier particles falling to the bottom of a mixture) causes uneven mixing. Convection currents have the same uneven mixing effects. In Spacelab, both these effects vanish. The preparation of extraordinarily pure medical products, such as vaccines, is

▲ This mock-up of Spacelab shows the same twin-module, single-pallet arrangement that will be used on the first flight. The working lab is to be delivered to NASA in 1980 and will fly in space a year later on the 11th Shuttle mission.

▲ ▼ Inside the Vehicle Assembly Building (VAB) at Kennedy Space Center, a fully-furnished Spacelab is lifted out of its shipping container and gently eased into the Orbiter's cargo bay. Above and below are two views of the operation.

A Laboratory in Space

Funding of Spacelab by ESA member nations

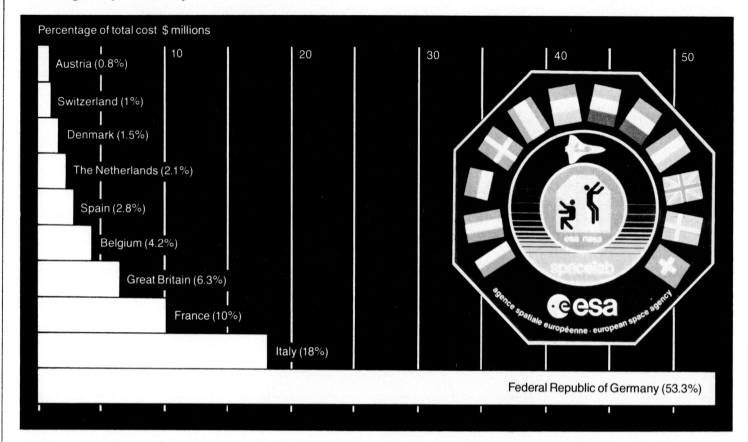

Percentage of total cost $ millions

Austria (0.8%)
Switzerland (1%)
Denmark (1.5%)
The Netherlands (2.1%)
Spain (2.8%)
Belgium (4.2%)
Great Britain (6.3%)
France (10%)
Italy (18%)
Federal Republic of Germany (53.3%)

10   20   30   40   50

agence spatiale européenne · european space agency
·e·esa

one area that will benefit greatly from these conditions.

A further advantage of space is that liquids float freely. Contamination caused by processing substances in containers can be overcome by doing away with them altogether and allowing the processes to take place in the open. Extremely pure glass for use in lenses and lasers can be made in this way, as can large, flawless crystals for electronic circuits.

A variety of objects, including ball bearings, filters, lamp filaments, and magnetic materials, are made by a process known as sintering. This involves fusing together powdered forms of different substances at very high temperatures. On Earth, sintered materials have a tendency to collapse. In space, the problem is readily overcome.

Another process with great potential in space involves the use of electrical charges to separate liquid mixtures. Experiments conducted during the Apollo-Soyuz programme showed that it may be seven times more efficient to separate kidney cells from urokinase, one of their products, in space than on Earth. Urokinase is the only known substance that will dissolve blood clots once they have formed. It is used to treat the victims of heart attacks, strokes, and vein inflammations.

Space research

A few examples of the experiments that will be con-ducted in Spacelab are listed below:

Multipurpose Fluids Experiment System (MFES) This consists of optical equipment that allows payload specialists to watch crystals growing from solutions and to see how the conditions of space affect their shape and structure.

Atmospheric Cloud Physics Laboratory (ACPL) The ways in which rain and snow form are still some of the least understood aspects of meteorology. The ACPL mini-laboratory has three cloud chambers in which cloud formation, the growth of water droplets, and the way in which electrical charges affect them, can be examined without interference from sedimentation and air turbulence.

Drop Dynamics Module (DDM) Drops of liquid will be injected into a chamber and positioned by means of sound waves. They can be spun or oscillated until they fly apart. Such experiments with the behaviour of fluids may reveal startling new information about the way in which everything from stars to atomic nuclei behave.

Vestibular Function Research (VFR) One of the major problems of spaceflight is nausea caused by weightless conditions. The VFR tests consist of four frogs in water in a centrifuge, where the 'gravity' they experience can be varied. Frogs and humans have similar inner ears that regulate balance. Their normal operation is disturbed when gravity is altered, leading to nausea and disorientation.

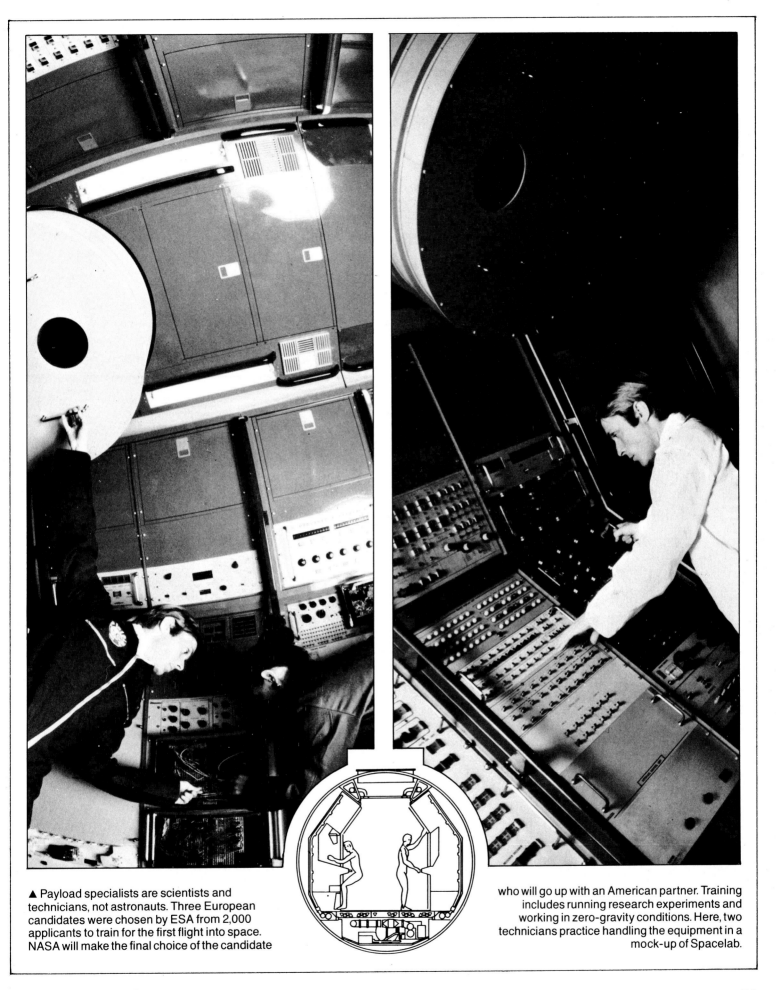

▲ Payload specialists are scientists and technicians, not astronauts. Three European candidates were chosen by ESA from 2,000 applicants to train for the first flight into space. NASA will make the final choice of the candidate who will go up with an American partner. Training includes running research experiments and working in zero-gravity conditions. Here, two technicians practice handling the equipment in a mock-up of Spacelab.

A Laboratory in Space

By the mid-1980s, a small fleet of Shuttles – based at Cape Canaveral, Florida and at Vandenberg, California – should be making dozens of flights a year.

A 'typical' Shuttle mission is hard to define inasmuch as every mission has different specific tasks to perform. The exact nature of any mission depends on the payload that is being carried. Yet though the details may vary the general sequence of events, from the launch to manoeuvring in orbit and re-entry, does not change very much from one flight to the next.

A typical mission lasts for at least seven days and may extend for as long as thirty. During this time, the Shuttle may function as a 'transport van', delivering and collecting satellites from space. On other occasions it may operate as a scientific research station, or even as a small-scale factory for processing various materials.

The start of a mission begins long before the dramatic last ten seconds of countdown, for the success of a flight depends to a great extent on early planning and preparation. Indeed, it is arguable that the most important advance in technology to come from the space programme is in the management methods devised to coordinate millions of operations in progress simultaneously all across the world.

The manufacture of the Orbiters is a classic case. The crew compartment and forward fuselage are made in Downey, California; the mid-sections in San Diego; the payload bay doors in Tulsa, Oklahoma; the wings in New York State; the wing leading edges in Texas; engines in Sacremento and so on. It is a miracle of management control that all these bits and pieces are produced in time.

In one sense, a mission can be said to begin when the main

▲ The loaded Orbiter is hoisted into position to be mated with an External Tank and a pair of Solid Rocket Boosters inside the huge assembly building at Kennedy Space Center. A giant crawler will trundle it out to the launch pad.

◄ ► With its five engines firing at once, the lift-off of the Shuttle is in every way as spectacular as that of the Saturn V which drew up to a million spectators to a launch. The three Orbiter motors burn with a pale yellow plume, the SRBs leave a smoky trail.

▲ The SRBs each generate 1.2 million kg of thrust. They separate at 44,200 m after a 123-second burn. Explosive bolts cut the struts and eight rocket motors fling the SRBs clear of the ET and Orbiter.

▼ The spent boosters parachute back and splash down in the sea at 94 km/h. On impact, beacons and flashing lights signal their location. The reusable rocket casings are good for 20 launches.

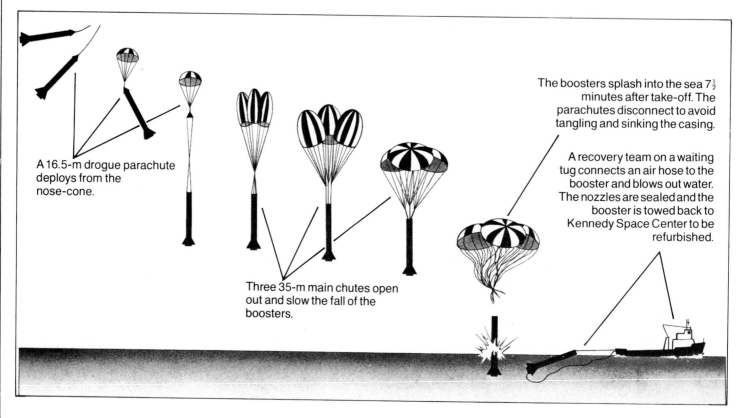

A 16.5-m drogue parachute deploys from the nose-cone.

Three 35-m main chutes open out and slow the fall of the boosters.

The boosters splash into the sea 7½ minutes after take-off. The parachutes disconnect to avoid tangling and sinking the casing.

A recovery team on a waiting tug connects an air hose to the booster and blows out water. The nozzles are sealed and the booster is towed back to Kennedy Space Center to be refurbished.

The Space Shuttle Handbook

▲ The 47-m long External Tank is jettisoned ten seconds after the main engines stop burning. It tumbles away from the Orbiter to break up during re-entry and scatter in the sea.

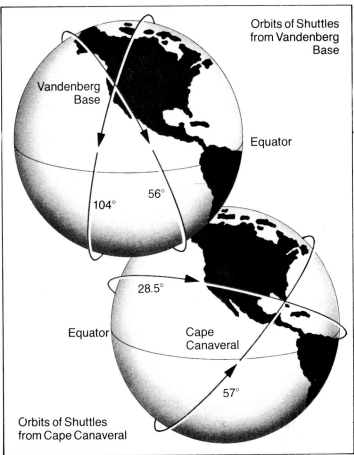

Orbits of Shuttles from Vandenberg Base

Vandenberg Base

Equator

104°

56°

28.5°

Equator

Cape Canaveral

57°

Orbits of Shuttles from Cape Canaveral

▲ Shuttle flights from Cape Canaveral will enter orbits between 28.5° and 57° to the Equator. Those from Vandenberg will orbit between 56° and 104°; limits that ensure that the SRBs and ET do not fall on land.

parts of the Shuttle arrive at the Kennedy Space Center. The Solid Rocket Boosters (SRBs) come by rail. The Orbiter arrives on top of a 747 Jumbo jet and the immense External Tank (ET) is delivered by barge from New Orleans. These three components are integrated in the Vehicle Assembly Building (VAB) – it became famous to the public during the Apollo moon shots as the 'biggest man-made structure in the world'. It is over 160 m tall, and more than four buildings the size of the United Nations in New York could fit inside it. The Shuttle is assembled in the VAB on a mobile platform that weighs over 4,800 tonnes. The whole assembly is then towed to the launch pad by a 3,000-tonne crawler with a crew of 15.

Loading the Shuttle

The way in which the payload is stowed in the Orbiter depends on its size. Usually, it is installed in the cargo hold before the Shuttle is taken to the launch site. The large Spacelab, for example, is gently hoisted into the bay by crane. However, it is still possible to load equipment while the vehicle is on the launch pad.

There is a fundamental difference between former generations of spacecraft and the Shuttle. In the past, vehicles were in a state of virtual isolation, save for the fuelling operation and boarding of the astronauts, once they had been taken to the launch pad. With the Shuttle, this is not the case – thus last minute loading is possible and the entire launch procedure is speeded up.

Cargo sizes

The amount of cargo that can be stowed is influenced by the launch site, either Kennedy Space Center or Vandenberg Air Force Base, and also by the orbit chosen for the flight. Payloads as heavy as 30,000 kg can be orbited if launched due east from Cape Canaveral. On the other hand, the Shuttle can only ferry 15,000 to 18,000 kg of cargo to high inclination orbits from Vandenberg.

Orbits will vary from 150 km to 900 km above the Earth, but most missions will take place around a height of 215 km. Lower orbits can be attained with less fuel. Since the trade-off is between payload and fuel, the more fuel it takes

continued on P. 60

FLYING THE ORBITER: A PILOT'S EYE-VIEW

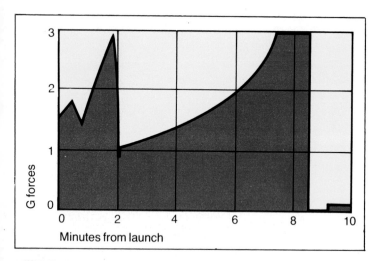

T he cockpit of the Orbiter looks much the same as the flight deck of a modern jetliner. However, there is one major difference. Unlike an airliner's crew, who line the craft up on the runway whilst in normal sitting position, the crew of the Shuttle lie on their backs facing straight up. The pilot's view through the flight deck windscreens is of nothing but sky.

The commander sits on the left, the pilot on the right. Each has a T-shaped group of instruments in front of him that contains an airspeed indicator, attitude display, altimeter and direction indicator. Further information is displayed on small TV screens in the centre of the instrument panel.

The commander's and the pilot's feet rest on aircraft-style rudder pedals. Pushing right turns the tail rudder right, pushing left turns it left. Between their legs, each has a pistol-grip control column. This has a dual purpose. In orbit it controls the Orbiter's attitude by firing clusters of thrusters. By pushing the column forward, the craft gently pitches nose down; by pulling it back it pitches nose up. A swing to either side makes the Orbiter roll in the same direction.

The second purpose of the control column is to deflect the Orbiter's aerodynamic control surfaces during re-entry and landing manoeuvres. There are five in all; two elevons on each wing and a body flap immediately beneath the nozzles of the main engines.

The elevons control roll and pitch in the atmosphere. By pulling the control stick back, all the elevons tilt up and cause the nose to pitch up. A downward push causes the opposite to happen. A flick left or right makes the craft roll in the same direction. The body flap has two purposes; it protects the engine nozzles from the fiery heat of re-entry and it assists the elevons to control the craft.

Next to the left hand of both the pilot and the commander is a lever that controls engine thrust during the launch and thus provides a measure of manual control to an otherwise automatic procedure. During the approach to touchdown it also serves as the airbrake control causing the rudder to split and open out into the airstream as two flaps that slow the Orbiter down.

Fly-by-wire

All the Orbiter's controls are of the 'fly-by-wire' type. Instead of the control stick pulling cables to adjust the elevons, its movements send electrical signals to the computers of the Flight Control System (FCS). The FCS receives the signals, then fires off an appropriate set of directions that move the elevons, rudder or thrusters.

The pilot has three control alternatives which can be selected by pressing buttons on the control panel. The first is AUTO. When selected, the craft flies itself and the crew become little more than passengers. Control Stick Steering (CSS) is the second alternative and enables the crew to fly the Orbiter using the fly-by-wire control system. Computers co-ordinate all the pilot's actions. Direct Control (DIR) is the third alternative. It overrides the computers. According to pilots who have trained on the flight simulator at Johnson Space Center, handling the Orbiter without computers is a very tricky task. At one stage during re-entry, for example, the aerodynamic forces of hypersonic flight reverse the effect of the elevons so that a command to roll left results in a roll to the right. In AUTO or CSS, the computers take automatic correcting steps. In DIR, a pilot must compensate for this himself.

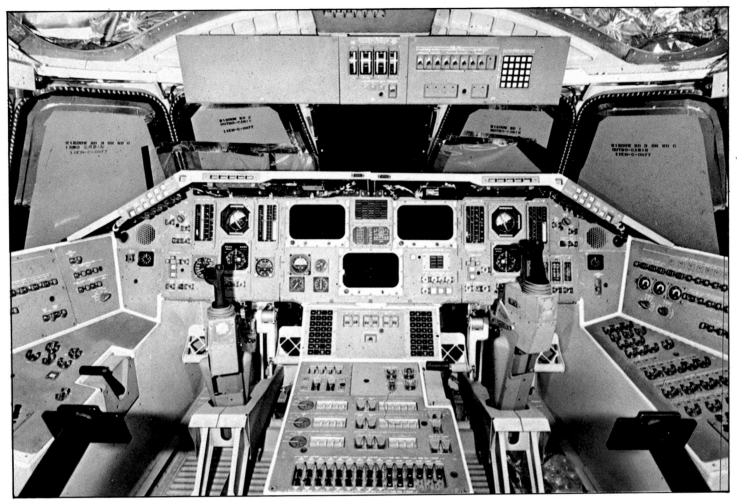

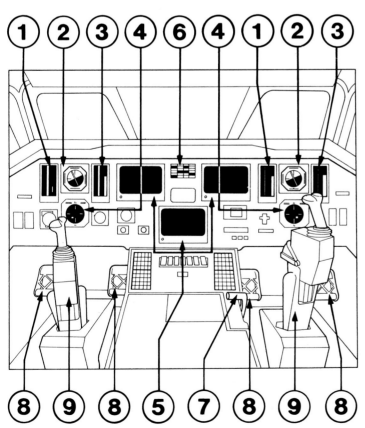

◀ ▲ The picture, above, is of the Orbiter's flight deck. The diagram, left, picks out the main instruments, most of which are duplicated so that the pilot and commander can each fly the craft. The instruments show: airspeed (1), attitude (2), altitude (3), direction (4). The TV screens (5) display additional computer information. Caution and warning lights are located on one panel (6). The controls for flying the Shuttle are the engine thrust and airbrake lever (7), the rudder pedals (8), and pistol-grip steering columns (9).

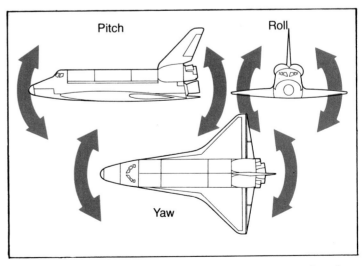

▲ Like any normal plane, the Orbiter is steered in the atmosphere by playing off the combined forces of pitch, roll and yaw. Any change in these causes the craft to turn, dive and alter speed.

The RMS robot arm

The Remote Manipulator System (RMS) is built in Canada. It is installed in the cargo bay alongside the payload. It can place and retrieve objects in space. The controller can follow the arm's movements through rear windows of the flight deck or by watching TV monitor screens. A TV camera is mounted on the RMS wrist.

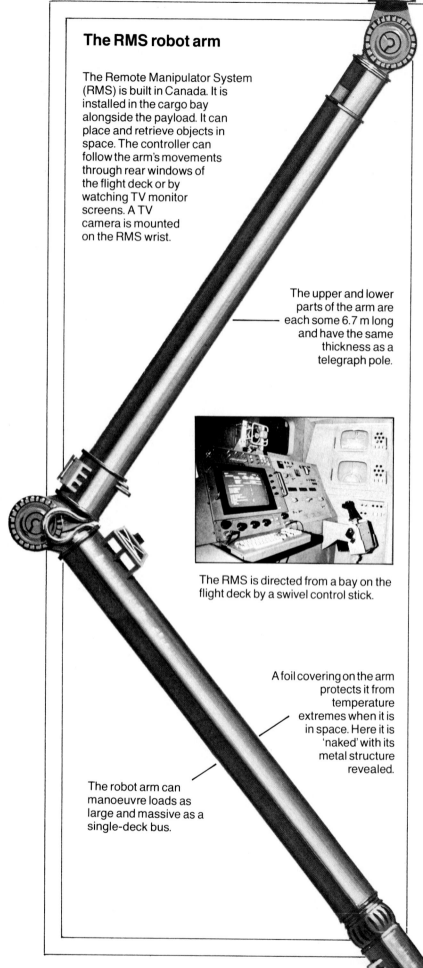

The upper and lower parts of the arm are each some 6.7 m long and have the same thickness as a telegraph pole.

The RMS is directed from a bay on the flight deck by a swivel control stick.

A foil covering on the arm protects it from temperature extremes when it is in space. Here it is 'naked' with its metal structure revealed.

The robot arm can manoeuvre loads as large and massive as a single-deck bus.

▲ Satellites are deployed from the payload bay and retrieved from space by means of a robot arm. The arm is manipulated by remote control from the aft flight deck of the Orbiter.

to get into orbit, the less cargo can be carried on board.

At lift-off, the two SRBs and the three main engines of the Orbiter fire simultaneously. The characteristics of solid-fuel rockets allow for a design with a slightly faster lift-off than earlier breeds of manned spacecraft. The ensuing acceleration is steadier and the maximum gravitational load is no more than 3 g, compared to 7 g and 8 g on Apollo flights.

After two minutes, and the consumption of some 900,000 kg of propellant, the Shuttle is 44 km high. At this point the solid fuel is used up and both SRBs separate. They descend by parachute; there are three main parachutes for each booster, each 35 m in diameter. The boosters fall into the sea about 220 km downrange.

Like so many aspects of the Shuttle programme, the recovery of the SRBs is an innovation. A team of divers plugs the exhaust nozzle of each rocket casing so that compressed air can be blown in to keep it afloat. The parachutes are reeled aboard the recovery vessels and two tugs tow the boosters back to Cape Canaveral. There, the centre sections that house the propellants are removed and sent back to the manufacturer for refilling. The nose cone and nozzle sections are refurbished on site.

Jettisoning the fuel tank

Fuel from the external tank continues to drive the Orbiter's three main engines until just prior to attaining orbital height and speed. Then the empty tank is jettisoned and drops back, breaking up and scattering into the Indian

Ocean some 16,000 km downrange.

Compared with the brute force of lift-off, the job of manoeuvring the craft into orbit and correcting the flight path is a delicate procedure. It is carried out by the Orbital Manoeuvring Subsystem (OMS) engines. These motors are also used to change orbits, manoeuvre to rendezvous with satellites, and to leave orbit for the return trip home.

Manoeuvres in space are also conducted with an array of little motors that keep the Orbiter in a predetermined attitude. This is known as the Reaction Control Subsystem (RCS); it consists of 38 bipropellant primary thrusters and six tiny vernier thrusters. Control sticks between the pilots' legs are used to flick the motors on and off for pitch-and-roll movements, while conventional rudder pedals are used for coping with yaw.

At work in space

Once the craft is in orbit, the commander and pilot move from the forward part of the flight deck to another panel of instruments along the rear wall. Mounted on the panel are hand controls to correct the vehicle's attitude, but in the main the equipment is for operating the Orbiter as a space station and working laboratory. The instrument panel includes controls for opening the doors of the cargo bay and for supplying power to the payload handling system.

Only when the vehicle is safely in orbit do the mission specialists and payload specialists begin their work. Payload specialists will not be used for the first five or six flights, when the most rigorous testing of the Shuttle system is to occur and the first trials to deploy and recover packages from the cargo bay are to be conducted.

The first Shuttle mission will carry no payload. It is a three-day test flight in a 278-km orbit. Throughout the mission, the Orbiter will fly in a 'barbecue' mode, rotating slowly to reduce thermal stress while it is being tested. At the end of the flight, a 200-second engine burn will transfer it to a re-entry orbit that will take it back to a touchdown at Edwards Air Force Base in California.

The second mission is a five-day one that continues the testing of the Shuttle system. The third mission is to be a seven-day flight to try out a payload and the robot arm that handles its deployment. The fourth flight is the last test mission.

By its very nature, a reusable space transportation system will have to be highly flexible. Because the Shuttle is devised to include both civilian and military operations, this means that no two flights are the same. Nevertheless, some procedures will be 'typical' of all missions. For example, a continuous watch over the power systems, environmental

▲ Not all satellites or upper stage rockets need a robot arm to be deployed. A yoke mounting can be used that swivels up from the cargo bay to release its payload into space.

controls and the guidance and navigation systems must be maintained. This involves a small mountain of paperwork. Every move of a spaceflight is rehearsed beforehand and listed on a flight plan hundreds of pages thick. The Shuttlenauts will have little time to sit twiddling their thumbs.

Keeping in touch

Communications links between ground control and the Orbiter carry voice, television, spacecraft and payload information. A very elaborate tracking network has been designed for this purpose. In the low Earth orbits where the flights are conducted, the Orbiter frequently loses radio contact with ground tracking stations. To maintain continuous links, therefore, special Tracking and Data Relay Satellites (TDRS) have been built. They are parked in geostationary orbits further out in space. There they can receive signals from the Orbiter and relay them back to Earth. Such a data handling procedure is highly complex. It calls for a big advance in the types of programmes needed by the on-board computers and for a considerable increase in the amount of computer operation by the crew.

Aiming the Orbiter

It must be remembered that for the most accurate scientific work, the Orbiter must be able to achieve and maintain any desired attitude. Special sensors allow this to be done by keeping the spacecraft pointing toward selected celestial or ground targets with an accuracy of plus or minus 0.5 degrees. A more complicated arrangement embracing the craft's guidance, navigation and control system, together with the Instrument Pointing System (IPS) in the payload bay, can increase this accuracy to plus or minus 0.1 degrees.

The return journey

When a mission is over, the Orbiter is positioned so it is facing away from its direction of travel and the OMS engines are fired to brake the craft into a re-entry trajectory. The return is a longer operation than the launch into orbit. The de-orbit burn causes the vehicle to lose height gently at first. Atmospheric re-entry begins at a height of about 140 km. At this point the craft is moving at 28,100 km/h. The effect of atmospheric friction and braking is not felt until 30 minutes later, at 130 km above the Earth. As the spacecraft glides through the upper atmosphere, the forces of deceleration build up, rising to 1.5 g over the next 20 minutes and remaining at that level for a similar amount of time. From 21,000 m to the final approach height of 3,000 m, the Orbiter's speed stabilises at 540 km/h. This part of the flight

▲ At the end of a mission the Orbiter positions itself tail-on to its direction of travel. It brakes by firing the OMS engines to slow its speed and begins to drop back toward Earth.

takes some $3\frac{1}{2}$ minutes. The normal landing sequence is fully automatic and the speed at touchdown is about 350 km/h.

Within an hour of landing on the 4,500-m runway, the Orbiter is towed to a workshop next to the VAB called the Orbiter Processing Facility. There the payload and remaining propellant are removed and servicing is carried out on the thermal insulation, the engines and the rest of the craft. Then it is moved to the VAB to be stacked with another External Tank and a pair of Solid Rocket Boosters and readied for the next flight.

Turn-around

There is one overwhelming difference between this procedure and the former process of attaching an Apollo spacecraft to the top of a Saturn V rocket. In the past, the preparation time for an Apollo launch was measured in months and the only part to return, the capsule containing the astronauts, ended up in one of the space museums. In contrast, the conversation at the Kennedy Space Center today is about turn-around times measured in days and life expectancies of 100 missions for each vehicle. Ground staff are geared to refurbishing an Orbiter in a week, integrating it with all the other components into a Shuttle configuration in three days, and moving the vehicle to the launch pad for another two days of checkout. Thus, landing to relaunching can be completed within two weeks (160 working hours).

Behind every Shuttle flight is the same calibre of highly trained ground staff, working 24 hours round the clock, as

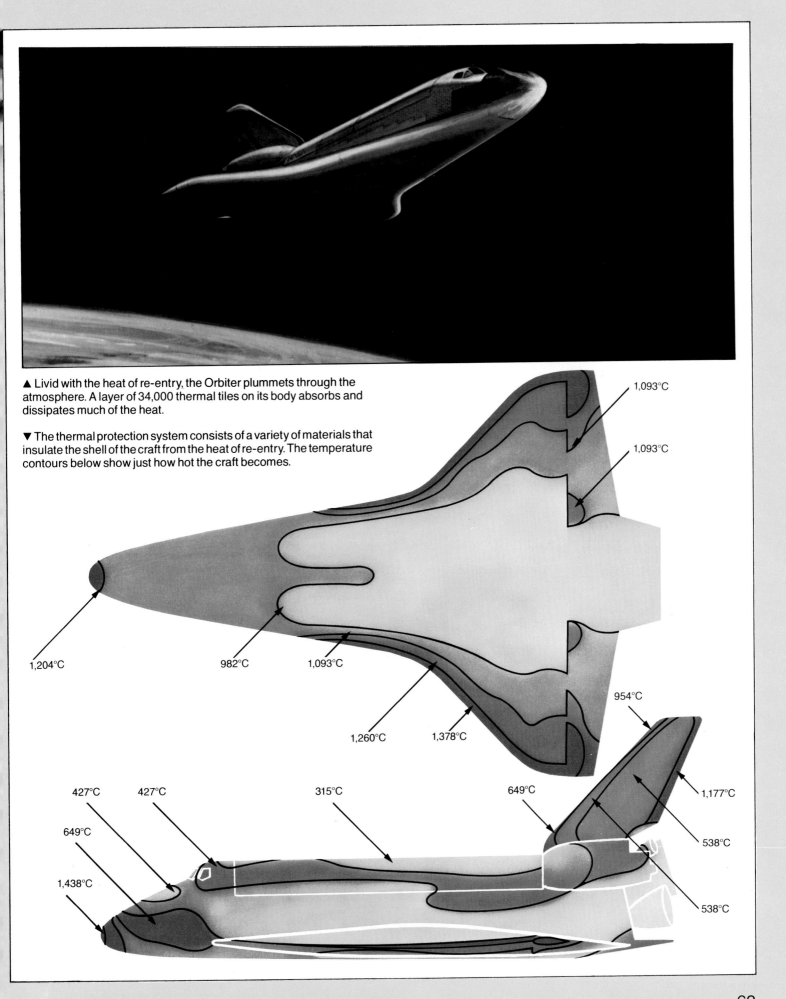

▲ Livid with the heat of re-entry, the Orbiter plummets through the atmosphere. A layer of 34,000 thermal tiles on its body absorbs and dissipates much of the heat.

▼ The thermal protection system consists of a variety of materials that insulate the shell of the craft from the heat of re-entry. The temperature contours below show just how hot the craft becomes.

1,093°C

1,093°C

1,204°C

982°C

1,093°C

1,260°C

1,378°C

954°C

1,177°C

538°C

649°C

315°C

427°C

427°C

649°C

1,438°C

538°C

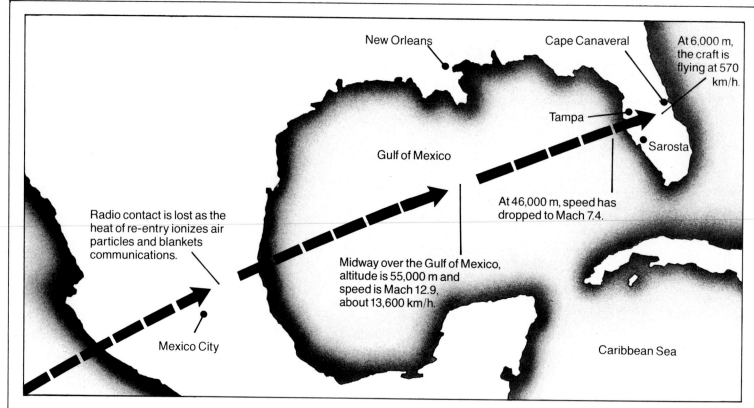

New Orleans

Cape Canaveral

At 6,000 m, the craft is flying at 570 km/h.

Tampa

Sarosta

Gulf of Mexico

At 46,000 m, speed has dropped to Mach 7.4.

Radio contact is lost as the heat of re-entry ionizes air particles and blankets communications.

Midway over the Gulf of Mexico, altitude is 55,000 m and speed is Mach 12.9, about 13,600 km/h.

Mexico City

Caribbean Sea

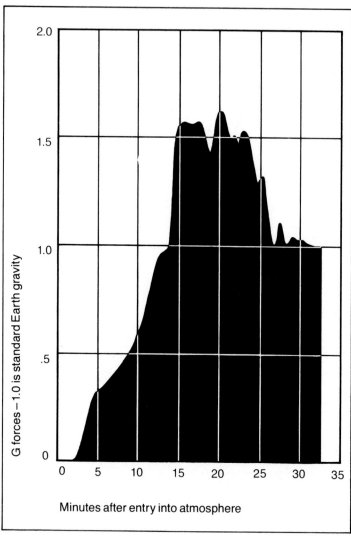

G forces – 1.0 is standard Earth gravity

Minutes after entry into atmosphere

▲ At 122,000 m, the Orbiter enters the atmosphere. It remains in a communications blackout until about 39,500 m. At 21,000 m, still travelling at 1,645 km/h, the TAEM computer programme begins to guide it onto its final approach and landing path. At 3,000 m, and a speed of 540 km/h, the approach to touchdown begins.

◀ As the graph shows, the gravitational forces encountered during re-entry are about half those of launching. They decrease rapidly once the craft begins normal atmospheric flight.

with all previous manned flights. In the same way as before, the launch sequence is handled by the Kennedy Space Center. During the orbital stage of the mission, control passes to Johnson Space Center in Houston. It returns to Kennedy for the landing.

Mission control

In one sense the Shuttle is a substantially slimmed down operation involving only 50,000 NASA, government and private industry employees, as opposed to the 200,000 of the fully geared up Apollo-Moon programme. Nevertheless, the drop in manpower does not bear a direct relationship to the drop in the programme's budget.

In the past, the Saturn rocket with its Apollo capsule on top, was at the launch site months before the lift-off. And only three or four launches a year were feasible, as against the 60 planned for the Shuttle by the mid-1980s.

The countdown control for the Saturn V needed a resident team of 500 people in the firing complex, with an extra 300 in the immediate weeks before the launch. For the Shuttle, a new procedure has been devised. Dubbed the

▲ Just minutes from touchdown, the Orbiter swoops over the sea, making its final course corrections before landing, in this case, on the runway at Vandenberg Base in California.

Launch Processing System, it developed from the earlier experience of the moon programme. This new method reduces the number of personnel in the firing complex to 50, including the specialists needed to monitor the circuits and instruments connected with the payload. A more sophisticated and highly automated system of data management and checking accounts for the change. In all, a ground staff of 10,000 at Kennedy Space Center can meet the needs of 60 Shuttle launches a year, compared to the army of 25,000 for a maximum of four Saturns.

Comparable savings in money and personnel have also been made with the launch pad. Earlier vehicles all needed a complex gantry tower that supplied essential services. It included a number of arms that embraced the rocket, and

that swung back during the last moments of countdown. It was absolutely essential that they were reliable, particularly those that stayed in place until engine ignition had started – failure would have been catastrophic for the spacecraft. Millions of dollars a year had to be spent on testing and maintaining the gantries to ensure that these structures were totally reliable. All this equipment has disappeared with the Shuttle. There are no rigid attachments at the launch pad. The only connection from the service tower, which also houses the elevator that takes the crew to the flight deck, is a flexible dropline carrying electric power. Most of the monitoring services are by telecommunication radio links.

Together with the scaled-down nature of the programme, these advances in engineering design have contributed a great deal to cutting the cost of a flight from approximately $445 million for Apollo 15 to a maximum of $27 million for the Space Shuttle.

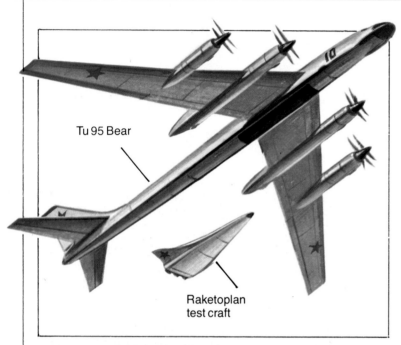

Tu 95 Bear

Raketoplan
test craft

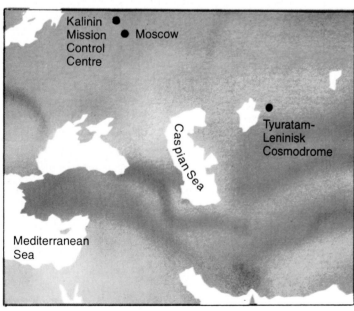

Kalinin
Mission
Control
Centre

Moscow

Tyuratam-
Leninisk
Cosmodrome

Caspian Sea

Mediterranean
Sea

▲ Test models of the Raketoplan have been air-dropped over Central Asia. It is thought to be the only other vehicle that is technologically a rival to the American Shuttle.

The Shuttle may be the first reusable spaceplane, but it is not the only vehicle that can put a cargo into space. A variety of other launchers offer stiff competition.

Three hundred kilometres above the flat steppes of Kazakhstan, to the east of the Aral Sea, an American spy satellite snapped a photograph that provided the first evidence of a direct competitor to the Space Shuttle. A Soviet One.

The satellite took photos of a new runway being built at the Tyuratam Cosmodrome that was particularly long and ideally suited for landings by the Soviet Union's answer to the Shuttle. The vehicle is known as Raketoplan (rocket plane). A prototype has been test-dropped from a Tu-95 Bear bomber in much the same way as *Enterprise,* though the American vehicle was perched on top of its mothercraft while the Russian one hung underneath.

The Russian mystery

Although information about the Russians' plans is sketchy, western experts believe that some versions of the Raketoplan may be able to replace the expendable Soyuz capsules as a means of re-supplying space stations of the Salyut type, as well as their projected successors. The Raketoplan would enable the Soviet Union to maintain an almost constant occupation of 'near-Earth' space, without having to build an enormous fleet of one-shot Soyuz supply vehicles and Progress space tankers.

The Russians themselves have released little information about their work. What is known is that the Raketoplan has delta wings and a tapering fuselage rather like an X-24A

▶ This scene of the Raketoplan is derived from a Russian painting and shows a future version rocketing into orbit. Its reusable launcher has already turned and begun to fly back to base.

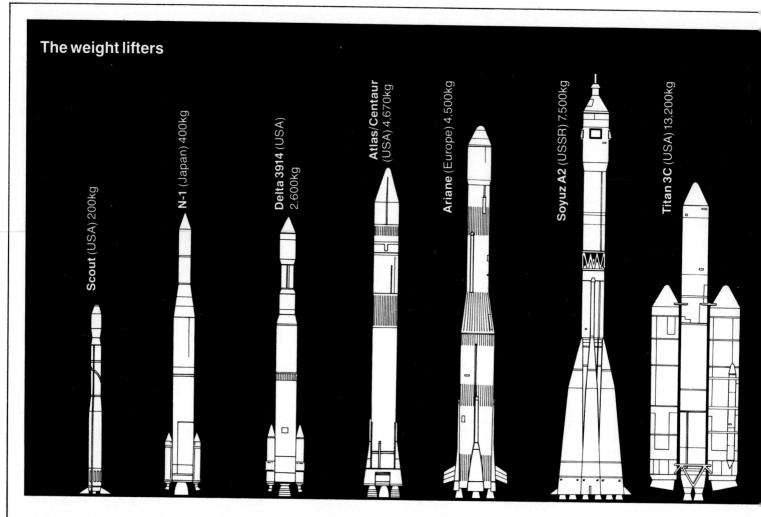

The weight lifters

Scout (USA) 200kg

N-1 (Japan) 400kg

Delta 3914 (USA) 2,600kg

Atlas/Centaur (USA) 4,670kg

Ariane (Europe) 4,500kg

Soyuz A2 (USSR) 7,500kg

Titan 3C (USA) 13,200kg

lifting body. Three powerful rocket motors are mounted at the rear. Radio Moscow has announced that it measures eight metres across and that its length is 60 m, although this figure includes the launcher – the craft itself is probably no more than ten metres long. If this is the case, the Raketoplan itself is somewhere between a Boeing Dyna-Soar and the Space Shuttle Orbiter in size. However, at 60 m it stands considerably higher on its launch pad than the Shuttle, which is only 47 m tall.

Experts in the United States who have examined the evidence believe that the combination of Raketoplan and throwaway launcher will be in use in the early 1980s to supply space stations with crews, fuel and other equipment. However, as far as is known, the Raketoplan is not thought to have as massive a payload bay as the Shuttle.

Still further in the future, it is believed that the Raketoplan will be mated with a reusable launcher. The combination, mounted piggyback, would take off from a runway like a normal aircraft. Having been flown to a suitable height, the Raketoplan would separate from its launcher and rocket into orbit while the launcher would fly back to base.

In the future, Shuttles and Raketoplans may blast off into orbit and wing their way back to Earth with the same regularity as commercial jets. But for the time being, the familiar single-shot launchers of the past two decades of spaceflight will continue to see service – at least for another ten years. Indeed, both Russia and America are still developing conventional rockets. Europe, China, Japan and India also have expendable launchers either in service or on the drawing board.

The space business

But if the Shuttle represents such a great technological step forward, then why are other launchers still being used? The reason is that the race to reap the commercial rewards of space is already well under way, and a veritable forest of rockets is being pressed into service to meet the demand. Countries all around the world are signing up to rent or buy communications satellites that can supply them with radio, telephone and television services. All need launchers. Not only does this mean business for work-hungry aerospace industries, it also provides a great deal of political prestige for those nations which have a thriving space programme. In addition, companies that can design and build rocket launchers can also supply the lucrative satellite side of the market; and along with satellites goes a wide range of

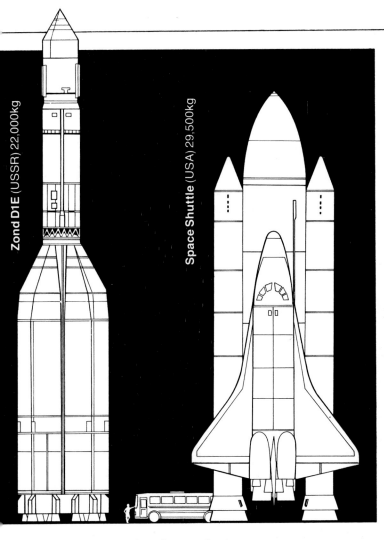

▲ To place satellites in orbits higher than its 1,200 km limit, the Shuttle will launch Solid Spinning Upper Stage (SSUS) rockets from its cargo bay. Up to four can be carried at a time. SSUS boosters can lift up to 2,000 kg into geostationary orbits. They are fired from an orbital height of 300 km.

◄ This chart compares the payload potential of various launch systems in reaching the same low Earth orbit.

tracking and broadcasting equipment.

Likewise, as the military uses of space expand, few countries will want to put their security into the hands of others. The United States and the Soviet Union would think twice about launching a foreign satellite that could shadow their own military installations. Nor would they wish to have their own secret cameras and communications equipment launched by a political rival.

The USA, USSR and China have all developed launchers from ballistic missiles that can be used to orbit military satellites as well as civilian ones. France plans to use Europe's Ariane rocket to launch a reconnaissance vehicle in the 1980s, and the privately-built rocket of the German OTRAG organisation is also being developed with the burgeoning satellite market in mind.

A third reason behind the continued development of throwaway rockets is that the Shuttle, which is designed primarily to put heavy loads into low orbit, is less cost competitive when it comes to injecting payloads into geostationary orbit 36,000 km above the Earth. Most communications satellites require this type of orbit, and at least 200 such payloads are planned for the 1980s.

The present generation of American, Soviet and Chinese launchers are based on vintage ballistic missiles first developed in the 1950s and 1960s. The Soviet launchers range from a small model based on the SS-4 Sandal ballistic missile to a giant rocket in the Saturn V class which, when perfected, may be able to put a staggering 150 tonnes into low Earth orbit. Thus far, however, it has shown a disturbing tendency to explode in flight. The Soviet Union has launched satellites for non-allied countries, notably France, but it is not in the commercial space business at the moment.

American launchers

In the United States, the Saturn series has been retired with the end of the Apollo and Skylab programmes. This leaves four families of launchers, based on the Thor, Atlas and Titan ballistic missiles and the small Scout rocket. All will continue to be used into the 1980s, with the Titan and Scout derivatives being probably the last to go to their well-deserved retirement.

A new version of the Titan, the 34D, has been developed to bridge the gap in American launch capability until the Shuttle becomes fully operational in the mid 1980s. It will also act as a back-up launcher in case problems crop up in the Shuttle programme. Titan 34Ds fired from Cape Canav-

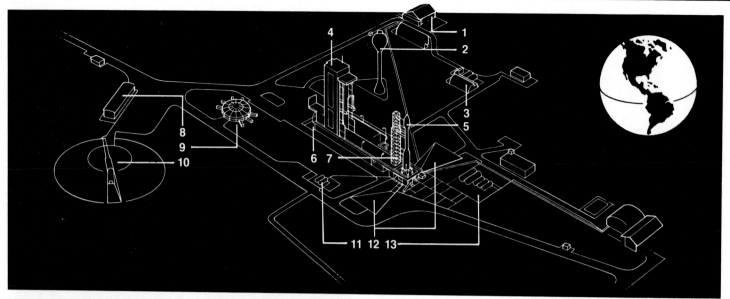

▲ A plan of the Kourou launch site in French Guiana shows: Fuel store (1), Water tower (2), Liquid hydrogen tanks (3), Service tower (4), Ariane (5), Helium store (6), Mast (7), Support services (8), Iced-water plant (9), Control bunker (10), Liquid oxygen store (11), Flame trench (12), Liquid nitrogen store (13).

◄ By early 1979, ESA had ordered five operational Ariane rockets and four test craft. The first trials were scheduled for late 1979; the first working missions a year later.

eral are to be fitted with an Inertial Upper Stage (IUS) rocket. It will enable the Titan to place more than 1,900 kg of payload into geostationary orbit, compared with the 1,430 kg limit of the earlier Titan 3C. Titan 34Ds fired from Vandenberg Base will not carry the IUS, but will be able to launch 14,900 kg into a circular orbit 185 km high, compared with the Titan 3C payload of 13,200 kg.

Medium-sized payloads have been traditionally launched by Atlas/Centaurs – the old Atlas intercontinental ballistic missile with a Centaur upper stage. The combination can inject 2,550 kg into low orbit, or 900 kg into geostationary orbit. The Atlas/Centaur will stay in service into the early 1980s. However, this slice of the market is squarely in Europe's sights, and here the Ariane rocket launcher competes head-on with its American counterpart.

In roughly the same class are the most recent versions of the old Thor ballistic missile, now known as the Delta 2914 and 3914. The latter, developed privately by McDonnell Douglas, can put 900 kg into geostationary orbit. The solid fuel Scout, which has four stages, is used to orbit smaller payloads.

The European challenge

Europe is currently waiting in the wings with a rival to the American medium-sized launchers. Ariane, now being developed for the European Space Agency, is designed to place satellites into geostationary orbits at a cost that

▲ A double-module OTRAG rocket being test-fired; this is the smallest version of the rocket. Extra modules can be fitted to enlarge it to almost any size. Simplicity and cost-cutting are the name of the OTRAG game. The propulsion system has only one moving part, the valve that controls fuel flow. The fuel is kerosene and nitric acid, a cheap though somewhat unreliable mixture.

The Competition

★In the spring of 1979, OTRAG's licence to conduct further tests was suspended by the government of Zaire following a growing number of international protests at its activities.

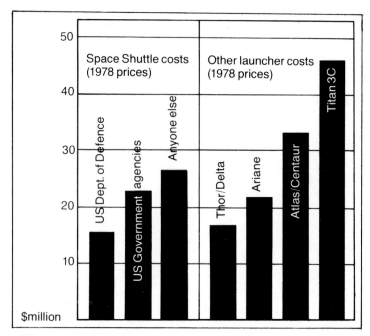

The chart shows bars for:
Space Shuttle costs (1978 prices): US Dept. of Defence, US Government agencies, Anyone else.
Other launcher costs (1978 prices): Thor/Delta, Ariane, Atlas/Centaur, Titan 3C.
$million (y-axis from 10 to 50)

▲ The cost of a single firing of the above launch systems does not take into account differences in payload size or orbit. Even so, the relative cheapness of the Shuttle is self-evident.

compares competitively with both the throwaway rockets of the USA and with the Shuttle. ESA's aims are fairly modest; Ariane is expected to capture anywhere from 16 to 62 of the 200 payloads expected to be orbited between 1980 and 1991.

Ariane is a three-stage rocket 48 m long and weighing 207 tonnes when fully laden. It can carry payloads up to 5.3 m in length and 3 m in diameter. The first versions are to lift 970 kg into geostationary orbit. Improvements are already in the pipeline and these will make the European launcher more attractive to potential customers. By 1983, the payload maximum will be several hundred kilograms greater.

The cost of Ariane

Prices will fluctuate for each flight according to the type of customer who is using the rocket, but ESA is determined to undercut its competitors. When bidding for the contract to launch the 950-kg Intelsat V satellite, in which it was successful, the European agency quoted a price of $20 million, a nine per cent discount from the basic launch price, compared with the $22 million quote for the Shuttle.

Ariane also has the ability to launch two payloads on a single flight. Thus, two satellites that might be orbited by a Thor/Delta rocket at $17 million a time, could both be launched with one Ariane for only $22 million. The Atlas/Centaur has a marginally higher performance than the European rocket, but it costs some $33 million per launch.

The Ariane project gains additional benefits from the use of Kourou in French Guiana as a launch site. Kourou is only

5° north of the equator – much nearer than Cape Canaveral – and payloads launched from here have the full benefit of the Earth's high rotation speed. This means that they need less fuel to reach a geostationary orbit.

Other competitors

China is also developing a civilian space launcher from a modified military ballistic missile. She has followed Europe's direction closely and her new rocket resembles Ariane. It has the same thrust (280 tonnes), weighs virtually the same (200 tonnes), and is capable of placing a 950-kg load into geostationary orbit. The first flight could take place as early as 1980.

Japan has similar ambitions. Its immediate work is with the N rocket, a design derived from the McDonnell Douglas Delta 3914. This launcher can put a 130-kg payload into geostationary orbit. The N-2 has bigger first-stage propellant tanks and uses extra strap-on boosters to lift heavier loads. Flights are due to begin in 1980.

A private launcher

At first glance, the most improbable of all the Shuttle's competitors is a rocket being developed by OTRAG, an organization financed by some 600 investors in Germany. OTRAG hopes to sign up the business which the big government space programmes will not handle for political reasons of one kind or another. The German company rejected the idea of using highly sophisticated propellants, as they require complex and expensive equipment to be handled. Instead, it has plumped for a kerosene-nitric acid propellant; old-fashioned perhaps but, OTRAG claims, perfectly suitable for its needs.

The OTRAG launcher is constructed from a cluster of pipes. The basic module consists of two propellant tanks and an engine. As many of these modules can be fastened together as necessary to provide the lifting power for a given payload. One tank in each module holds the kerosene, the other contains the nitric acid. The tanks are emptied by compressed air so there is no need for complex and expensive pumps to feed the fuel into the engines.

The engine of each module produces a maximum thrust of three tonnes. Two modules joined together form the smallest version of the rocket. By adding extra modules, the rocket can be tailored to reach different orbital heights and carry payloads of different sizes.

OTRAG hopes to have a giant 600-module rocket ready for firing by 1981. It should be able to launch 1,500 kg into a geostationary orbit. This performance is similar to that of the Atlas/Centaur and to early versions of Ariane.

▲ An OTRAG team tests the assembly of a fuel tank cluster in Germany. Later it is shipped to the launch site in Zaire, where OTRAG has leased a test range half the size of West Germany.

▶ The third OTRAG launch went wild after a few seconds of flight in June 1978, and veered off course to destroy itself.

▼ The launch site sits atop a 1,300 m high plateau in Shaba Province. A natural fortress, it offers a considerable degree of security in this politically troubled corner of Africa.

9 THE FUTURE

The exploitation of space has barely begun, yet already projects of a staggering scale have been envisaged. Their potential benefits are enormous, but so are their costs. The Shuttle, by virtue of making space accessible in a routine way, holds the key to their success.

▲ This Boeing concept of a satellite power station has panels of solar cells measuring 29 km by 6.5 km. It can generate 10,000 megawatts of electricity, some ten times the capacity of the biggest nuclear plants. The scene (right) shows the construction of one such power station. These structures could be operating by the year 2000.

We are no longer in the realm of science fiction when we hear talk of orbiting laboratories, space factories and communications platforms. The possibility that such projects will come into being in the not too distant future – certainly within our own lifetime – has become all the more likely because of the Shuttle, the transportation system that enables people to commute regularly between Earth and space.

One likely development will be the advent of power stations in space. While fuel supplies on Earth are by no means about to run out, they are becoming increasingly scarce and expensive. As they do, the search for new sources of energy becomes imperative. And so, even such seemingly far-fetched ideas as trapping solar energy in space and pumping it to Earth acquire a certain merit.

Power from space

A revolutionary idea for beating the fuel crisis by tapping the energy of sunlight in space has been proposed in the United States. It began with studies by the Arthur D. Little Corporation and has since been the subject of research by NASA laboratories, and major aerospace companies including Grumman, Rockwell and Boeing.

The advantages of orbiting power stations are obvious. Unhindered by bad weather or limited daylight, they can generate a steady stream of energy. Dr. Krafft Eriche, scientific advisor to Rockwell, estimated that a Satellite Solar Power Station (SSPS) could give as much pollution-free electricity in one year as can be obtained from 3,700 million barrels of oil – fully 50 per cent of the entire Middle East output exported in 1973.

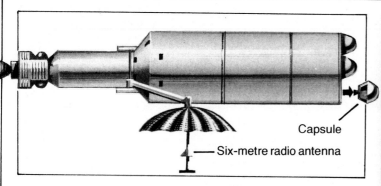

Capsule

Six-metre radio antenna

▲ ▼ The 'Big Bird' reconnaissance satellite can deliver pictures without leaving orbit. As it overflies Alaska, it ejects one of six capsules loaded with exposed film. The capsule fires its braking-rocket and plunges back to Earth to be snagged by a Hercules aircraft as it parachutes down near Hawaii.

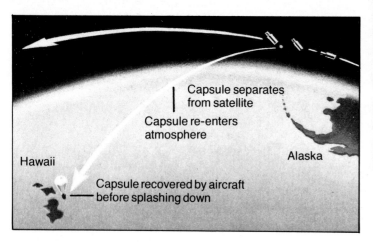

Capsule separates from satellite

Capsule re-enters atmosphere

Hawaii

Alaska

Capsule recovered by aircraft before splashing down

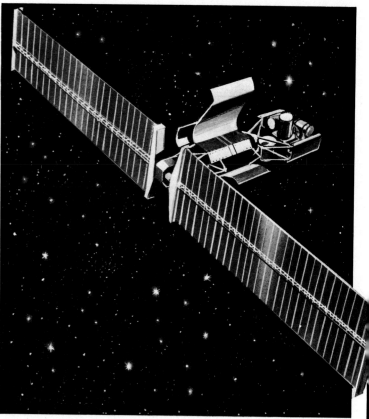

▲ The wing-like panels of this power satellite are solar cells. They can generate up to 25 kw of electricity for use by the Shuttle in order to extend its stay in orbit.

It remains to be seen how practicable such power stations would be as a means of lessening our dependence on oil. Certainly they would have to be immense structures many kilometres across in order to capture sufficient sunlight. Astronauts would have to build them in space using a fleet of Shuttles and Space Tugs to help with the assembly.

Such power stations would consist of giant arrays of solar cells; one NASA design measures eight kilometres by eight kilometres. The electricity generated in space would be converted into high energy microwaves and beamed back to Earth. There it would be transformed into direct current for distribution. The energy beams could be switched from country to country according to need.

However, such grandiose ideas are still several decades away. The most likely projects at the moment are more modest affairs.

Space war

The tremendous advances in the civilian uses of space have tended to blur the military origins of so many developments. The close collaboration between NASA and the Department of Defence to make the Shuttle programme a success is not always fully appreciated. The whole operation was designed so that fully one-third of the flights would be under military auspices. For this purpose, two out of the total fleet of five Orbiters are to be stationed at Vandenberg Air Force Base, from where up to 40 military-sponsored missions a year will originate.

Already we depend upon satellites for passing military communications around the world and for precise navigation by ships and aircraft. Reconnaissance satellites are vital to ensure that arms control agreements are not infringed, while others are able to give advance warning of attack by picking up the infra-red exhaust signatures of ballistic missiles.

Hunter-killer satellites

This increasing dependence on the 'high-ground' of space for military purposes has led the Soviet Union and America to develop interceptor satellites, or 'hunter-killers', that can destroy other satellites.

The problem with war in space is that any orbital vehicle, unless it is highly manoeuvrable, is a sitting duck for missiles or hunter-killer satellites. On the other hand, the rate at which defence satellites can now be launched means that a huge offensive would have to be staged to eliminate them.

▲ Satellites can be knocked out by energy beams fired from a great distance away. If not destroyed outright, their delicate sensors and other vital equipment are easily put out of operation. It appears that this could be done far more effectively from space by highly-manoeuvrable 'hunter-killer' satellites than by firing laser or particle beams up through the atmosphere.

It is conceivable that the Shuttle might one day operate as a 'space interceptor' by using its cargo hold as a weapons bay. Many roles are possible with a vehicle of this size, and it would be surprising if the US Air Force has not considered turning the Shuttle into a platform on which to mount laser weapons of the kind being studied for a new generation of supersonic bombers.

By virtue of being highly manoeuvrable, the Shuttle could also rendezvous with low-orbit 'enemy' reconnaissance satellites and literally pluck them out of orbit, tuck them into its cargo bay and fly home with the booty.

Space Tug

A major technical contribution by the military to the Space Shuttle Programme was the idea for an Inertial Upper Stage (IUS), or Space Tug. The requirement by the Air Force for a booster which could lift heavy reconnaissance satellites into orbit was responsible for this development.

Military satellites and other payloads will be carried into low orbit by the Shuttle. However, to boost them into operational orbits as high as 36,000 km, or even into interplanetary trajectories, requires a second space launch. Boeing are building solid fuel Space Tugs to fulfil these requirements. They come in a number of two and three-stage versions.

Space Tugs are to be launched from tilt tables in the Shuttle's cargo bay. They are separated by spring-ejectors to float free of the spacecraft. Then their guidance systems are aligned and they are fired into orbit by remote control.

Applications of the Space Tug are not limited to just military payloads. Weather, navigation and communications satellites that travel in geostationary orbits can all be launched by Space Tugs. They will also be vital in lifting large space structures into higher orbits.

Shuttle power-pack

Another small-scale development that stands a chance of being used in the foreseeable future is a solar energy module, which can either remain in the cargo bay or fly free. This power source could extend the Shuttle's stay in orbit.

First proposed by the Marshall Space Flight Center, this module, with its large extending solar 'wings', would supply continuous power to augment Shuttle resources and allow more extensive activities for the Shuttle-Spacelab systems.

The original idea was that the module's main structure would consist of the open rack of an Apollo Telescope

▲ The Space Telescope will be deployed from the Orbiter's cargo bay into orbit 480 km above Earth. Once in position, it will be operated by remote control from the ground. With no weather to hamper its operation, it will be able to detect objects 50 times fainter than ever seen from Earth, even by telescopes twice its size. It will be able to 'see' light from sources up to 14,000 million light years away.

The guidance system is fantastically accurate. It will aim the 14.3 m telescope to within 7/1000th of a second of arc – an accuracy that would enable a marksman on the ground to hit a penny-sized target 450 km away.

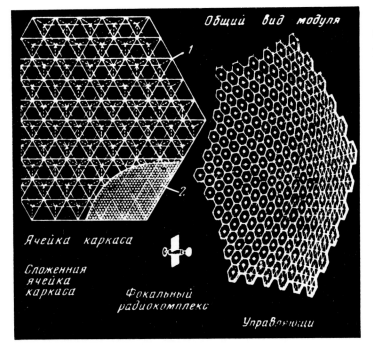

Общий вид модуля

1

2

Ячейка каркаса

Сложенния
ячейка
каркаса

Фокальный
радиокомплекс

Управляющи

▲ Calculations have shown that beehive constructions would be very promising for building space platforms. The Soviet Superscope, a huge ten-kilometre reflector, is assembled from thousands of honeycomb-shaped modules, arranged into any desired shape.

The telescope will remain in orbit for at least five years after its planned launch in 1983, transmitting the results of its findings back to Earth. Then it will be brought down by Shuttle for servicing before being re-launched. Solar panels along both sides will provide it with a constant four kilowatts of electric power during its stay in space.

Superscope

The Soviet Union, too, is looking into massive science projects in space. One structure now under study is a giant 'Superscope' which could revolutionise the study of the universe, top Russian scientists believe.

They envisage this radio-telescope to have a huge bowl-shaped reflector one to ten kilometres across. It will be assembled piece by piece from honeycomb modules 200 metres in diameter. Each module will have a four-metre wide reflecting cell at its centre; its tilt would be regulated by commands from an unmanned spacecraft positioned in front of the structure.

Apart from enlarging the scope of radio-astronomy in general, the Superscope could also be used to search for faint signals which may be reaching the Earth from other intelligent beings in the universe. Calculations show that this size reflector could theoretically pick up a one-megawatt transmitter ten to 100 light years away. A more powerful signal being directed at our solar system would be detectable from as far away as three thousand to ten thousand light years.

Superscopes would float high above the Earth – ideally at one of the Lunar Libration Points (384,000 km away) – to limit interference from our own broadcasts that might block out faint signals from space.

International co-operation

Lifting the parts into space for assembly would be a major problem, but according to one Russian commentator, "the general state of the art has developed to the point where the establishment of large complexes of this type should be feasible in the very near future." It was for convenience of transport that four metres was chosen as the basic size of the flat reflecting cells used in the design.

Studies of Superscopes have also been prepared in the USA by the Ames Research Center in California. Soviet scientists say they would welcome international co-operation on a project of this magnitude. Given the necessary political goodwill, we may one day see Soviet and American launch vehicles jointly ferry up the necessary components.

At the other end of the scale from the 'big science' projects

Mount (ATM), available from the Skylab programme – thus making use of previously developed space hardware – and two sets of solar arrays being developed for a new kind of propulsion system. Control gyros from the ATM would provide stability for the satellite and for the Orbiter when both were docked together.

The power module could later be attached to an External Tank of the Shuttle, which would act as a 'strong back' to support space construction activities of various kinds. NASA also regards the power module as a step towards a semi-permanent observatory that would include the large double module from Spacelab, and could be left in orbit between Shuttle visits.

Space science

One of the most exciting research projects in space is the Space Telescope. This is a free-floating optical laboratory some 14.3 m long and 4.7 m in diameter consisting of a telescope, a package of detecting and communications equipment and a power and propulsion system. Its primary mirror is 2.4 m across and compares favourably with the big 2.54 m telescope on top of Mt. Wilson in California.

From its lofty position in orbit, the Space Telescope will be able to 'look' seven times farther into space than any Earth-based model. Its resolving power is so fine that it would be able to distinguish a large-sized coin at a distance of 580 km.

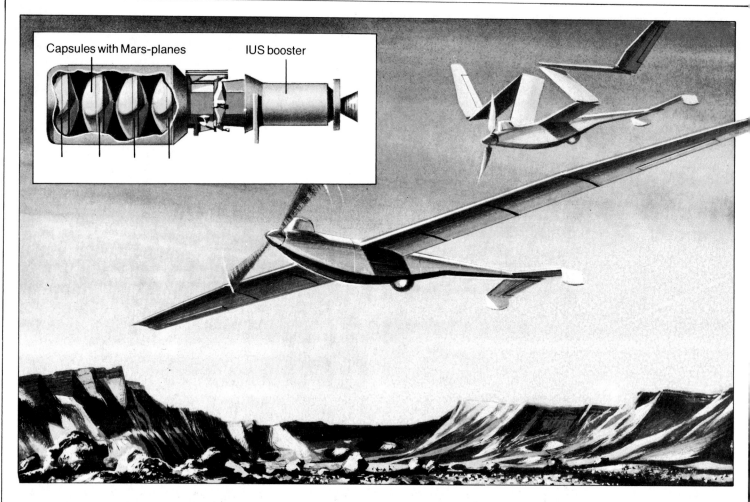

Capsules with Mars-planes   IUS booster

▲ Four Mars-planes stowed in capsules could ride atop a two-stage IUS rocket that would boost them from the Shuttle's low Earth orbit into a path that intercepts Mars. Once in a Mars orbit, the mothercraft would eject the four capsules. The robot plane would 'hatch' from its entry capsule at 7,500 m like a giant butterfly and begin a cruise through the thin atmosphere that could last up to 31 hours.

of the USA and the USSR are the hundreds of small scientific payloads that can be sent up by scientific organisations, colleges and schools that have paid their deposits for these 'getaway specials'. Experiments such as these can be taken into orbit by the Shuttle from as little as $3,200.

Interplanetary missions

Bold plans have also been made to use the Shuttle to launch interplanetary spacecraft from its cargo bay, using multi-stage rockets to propel them into space. Such robot explorers can study other planets either by flying past them, orbiting them, or sending a lander down to the surface.

One such mission, known as Project Galileo, calls for parachuting an instrument capsule into the dense, turbulent atmosphere of Jupiter in 1985; the project is due to get underway in 1982. The mothercraft will swing into a wide orbit around the giant planet, after releasing its instrument package, to conduct a year-long survey. It will also receive data from the falling capsule, boost its signal and radio it back to Earth.

After the instrument capsule has plunged into the Jovian atmosphere, it will gradually slow down until, at subsonic speed, its heat shield drops away and the capsule drifts down under a parachute. After some minutes, the parachute separates and the capsule falls freely until the enormous pressures encountered crush the shell.

Mapping Venus

Another NASA plan for the Shuttle is to put a radar-equipped satellite into orbit around Venus to penetrate its thick cloud cover and map surface features. The study will reveal the presence or absence of continents, ocean basins, mountain chains, rift valleys, craters and volcanoes.

Studies have also been made of the possibility of flying a space probe through Halley's Comet when it rounds the Sun in 1985-86. The probe would continue on its way to intercept another comet, Tempel 2.

What many scientists are waiting for is the chance to use the Shuttle to renew contact with Mars. One early idea was to have two Shuttles rendezvous in Earth-orbit, one carrying

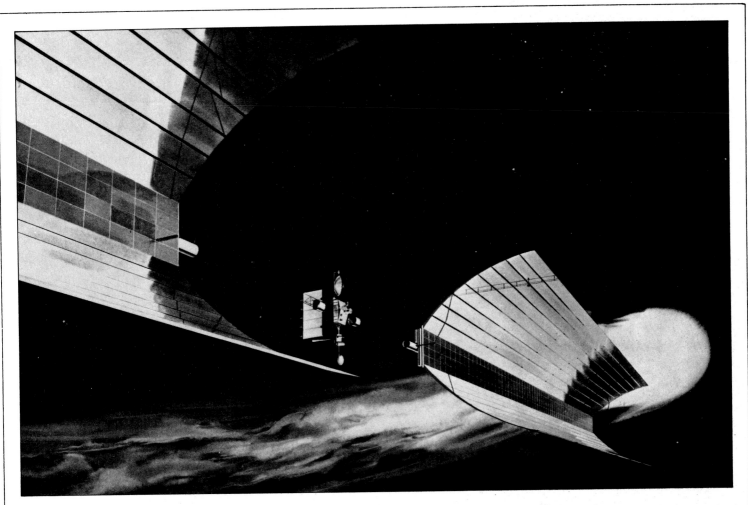

▲ An ion-drive spacecraft draws near to Halley's Comet on its way to intercepting it. This plan could be put into effect in 1986 when the comet next nears Earth. The reflector 'wings' of the craft are each 8 m wide and 60 m long. Their scooped-shape focuses sunlight on a bank of solar cells. These, in turn, provide the electric power to drive the mercury ion engines.

a Mars-lander and the other a propulsion unit. This would enable a heavier payload than ever before to be launched to the Red Planet.

One possible use of the larger payload capacity will be to land small tracked robots that can travel over the planet's surface and analyse the soil in different places.

The planes of Mars

An especially ingenious scheme is one that calls for releasing robot aeroplanes into the thin Martian air. They would make low-altitude reconnaissance flights to report on the planet's surface features.

These ultra-lightweight craft would have to fold up inside the stowage compartment of the entry capsule. One design envisages packing four such planes into separate capsules that are released one after another from the orbiting mothercraft. The orbiter also acts as a relay station, passing messages between the planes and Earth.

The entry sequence would take place as follows. A capsule drops away from the orbiter. After taking the brunt of the frictional heating, the aeroshell is slowed by parachutes to a speed of 216 km/h. then it splits open and the bottom falls away.

Still hanging beneath the parachute and the capsule top, the plane starts to unfold like a great ungainly insect coming to life. Its sectioned body hinges into place and the wings flip-flop open to their full spread of 21 m. A huge 4.5 m propeller opens out and comes to life. The dearth of free oxygen in the Martian atmosphere means that the engine cannot be petrol driven, so the designers have proposed using a motor driven by hydrazine. Immediately upon being released, the plane dives vertically, then pulls out to horizontal flight at an altitude of 5,700 m, to cruise along at 380 km/h.

Normally, this strange winged craft would carry out geological surveys at a height of some 15 km, but there is a version that could be programmed to fly low over the ground and even to land on its stilt-like legs. Scientists at the Jet Propulsion Laboratory in the USA have suggested that it be used as a delivery system for scientific packages, or to

collect samples for subsequent pick-up by robot landers.

Calculations suggest that a Mars aeroplane could carry a payload of 40-100 kg and still fly as far as 10,000 km.

Factories in space

The Shuttle does much more than simply expand the range of payloads that can be orbited around Earth or sent into deep space. It provides a working environment in which a tremendous variety of materials can be processed in the unique conditions found in space.

America and the Soviet Union have already carried small electric furnaces into orbit to study the prospects of processing materials. There is great promise, for example, in making high-purity crystals for the electronics industry, and ultra-pure vaccines and serums for medical treatments.

Foam steel

New kinds of ultra-lightweight alloys are also feasible. As one senior NASA official put it; "We'll develop super-materials which may revolutionise construction techniques on Earth. We may manufacture metal foams by injecting air into liquid metals much like yeast in bread dough, and let it solidify in weightlessness."

"A foamed steel with 87 per cent gas content would actually float on water like cork! Imagine what foamed steel would mean to an aircraft wing if it is as strong as stainless steel and as light as aluminium."

If the processing of materials in space turns out to be feasible on a large scale, the next step will be to build orbiting factories. It will take a host of ingenious building machines to assemble structures of this sort. Several have already been designed as 'accessories' to the Shuttle.

The Beam Builder

One, the Beam Builder, is a device that is being developed jointly by NASA and the Grumman Aerospace Company. Fitted into the spaceplane's cargo bay for transportation into space, the Beam Builder is designed to shape light-weight

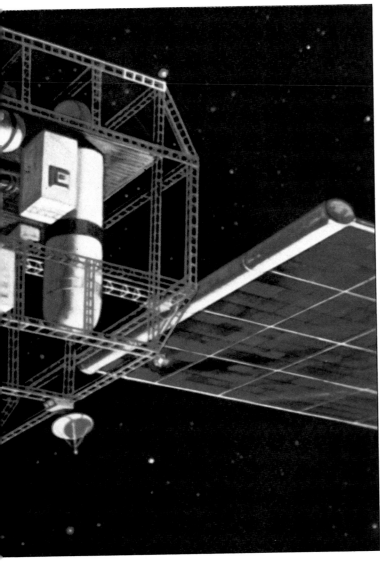

▲ The Public Service Platform is a project with tremendous potential. Among its many possible uses is the electronic distribution of mail for the entire USA. This array has 91 ten-metre dish antennae.

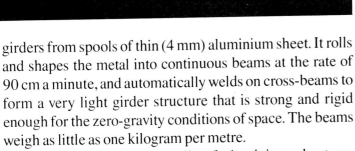

◄ The robot arm of an Orbiter plucks a load from the cargo bay and swings to position it in the huge space factory being assembled in low Earth orbit.

girders from spools of thin (4 mm) aluminium sheet. It rolls and shapes the metal into continuous beams at the rate of 90 cm a minute, and automatically welds on cross-beams to form a very light girder structure that is strong and rigid enough for the zero-gravity conditions of space. The beams weigh as little as one kilogram per metre.

A single load of three rolls of aluminium sheet are sufficient for the Beam Builder to produce a continuous girder one metre wide and 305 m long. In theory, such a machine could be easily reloaded and carry on churning out beams of unlimited length.

Space Spider

One novel building tool is the Teleoperator Space Spider. This could be a forerunner of future generations of intelligent building machines. The Space Spider is a self-contained system. It transforms spools of coded, pre-stamped building materials into complete assemblies in a single, integrated operation. As it works, it resembles

nothing so much as a spider spinning its web.

The Space Spider moves by a sprocket drive along the rim of a fixed core of material. As it crawls along, a spool of pre-stamped material is unrolled and attached. The process continues until the desired size is reached. Such a machine would be perfect for building the huge panels of a solar power satellite.

Space stations

Several studies by NASA's Marshall Space Flight Center have toyed with the idea of reusing the Shuttle's big External Tank as the core of various space structures. The 8.4 m diameter tank is big enough to make an ideal semi-permanent space station. At launching, the tank would be only part-filled with fuel while a separate compartment at the front end was fitted out as the crew's quarters. Here the crew would live while they assemble a big space structure.

Once the ET was parked in orbit, a second Shuttle would bring up an airlock module and a docking bay, as well as solar

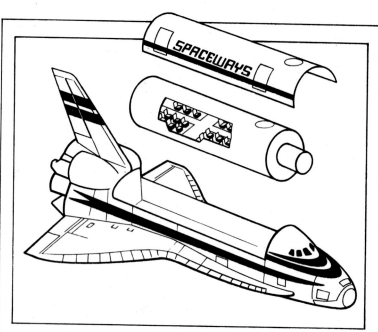

▲ The Orbiter could be transformed into a passenger-carrying craft shuttling between space stations and Earth. Here, a compartment that seats up to 74 people is slotted into the payload bay.

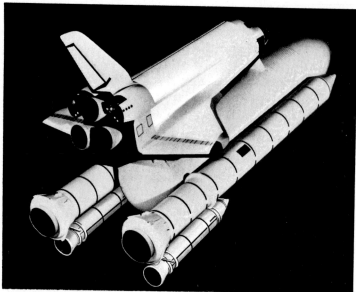

▲ One of the most likely ways to uprate the payload capacity of the Shuttle, and the orbits it can attain, is by adding strap-on boosters to the existing SRBs.

panels that would be unfurled and attached to the station to provide it with power.

When these units have been linked with the tank, the space station would be fitted out with enough supplies to support a crew of three for anything up to six months.

Similar proposals have been made to assemble a modest space station from separately-launched laboratory modules – each one similar in size and shape to Spacelab – but capable of being removed from the Shuttle's cargo bay to float in space.

Public Service Platforms

Another exciting development, at the moment forecast for the 1990s, is the Public Service Platform (PSB). This is a communications system that, it is envisaged, will float in geostationary orbit high above the Earth. The PSB consists of a grid of small high-powered antennae, each operating on a different frequency. It will measure as much as 300 metres across.

The PSB will be very different from present communications satellites that transmit telephone and telex services around the world and bring overseas events to our TV screens. Such systems require large ground antennae installations and powerful signalling, receiving and processing equipment, not to mention a further distribution system to relay the programme to individual consumers.

The Public Service Platform reverses this procedure. It takes the broadcasting equipment directly into orbit so that ground transmitters can, in turn, become slimmed down devices operating at low power. The PSB has the potential to

provide such services as wristwatch telephones, electronic mail sorting, educational television, disaster warning and search and rescue information, as well as countless others.

Building such a large structure calls for the method of assembling it in low Earth orbit and then lifting it into a high geostationary orbit.

Switching orbits

One example of an unmanned orbital transfer system is the Solar Electric Propulsion Stage (SEPS). It will employ a battery of ion motors – devices that convert electrical energy and a metallic propellant into a thrust force that consists of a stream of ionized particles. Typical propellants are mercury or caesium. The electrical power is supplied by big arrays of solar cells.

SEPS is particularly attractive for transferring large structures because it is highly efficient with its fuel and can apply a small amount of thrust over long periods, thus placing little strain on the structure being lifted.

Future Shuttles

One of the difficulties encountered in assembling giant structures in space is the problem of lifting huge quantities of building materials into orbit. Restrictions in the payload capacity of the present Shuttle will make it necessary to improve its performance in the not so distant future. And following from that, a second generation of heavy-lifting Shuttles is the next obvious step.

The cheapest solution is to start by modifying existing Shuttles to improve their performance. In the next few

▲ One development from the Shuttle is a freighter that can carry up to 135 tonnes into low orbit. The Orbiter is replaced by a huge cargo-cylinder attached to an engine pod. The four liquid-fuelled boosters have clamshell doors at their nozzles. These close to protect the boosters during the impact at sea. The two-booster version, inset at top, lifts up to 90 tonnes of freight.

▼ The giant kite-shaped reflectors of this solar power station stretch 24 km through space. They focus sunlight onto heat engines. The largest such satellites weigh 80,000 tonnes and are 150 km² in area.

▲ The heart of the solar power station is a group of spherical heat engines. They absorb the heat of the reflected sunlight and use it to generate the electricity that is beamed to Earth.

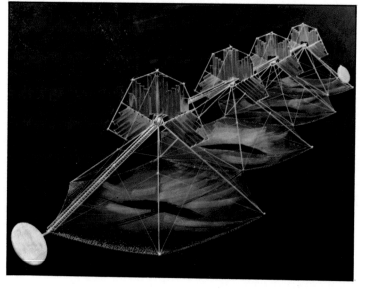

years, it will theoretically be possible to lift as much as 1,770 tonnes of freight into orbit per year by flying between 40 and 60 missions annually; although a figure half this size is much more realistic. This would be quite acceptable to support a space programme that consists primarily of semi-permanent orbiting stations, laboratories, satellite launches and interplanetary missions. But it comes nowhere near the figure that will be needed to build large structures.

For example, a fully-fledged Solar Power Station (SSPS) programme would require at least 30 giant satellites to collect reasonable quantities of solar energy. The Shuttle would have to carry something of the order of 125,000 tonnes of cargo a year to maintain such a level of activity.

Some five different classes of heavyweight Shuttles have been examined by NASA. Their payload capacity ranges

▲ The ground-based 'rectenna' is a large array of receivers that collect the microwave beams from space and convert them into direct current. It covers an area of some 96 km².

from 60 tonnes to more than 300 tonnes. As these craft are intended mainly for use as cargo freighters, they would be crewless; a consideration that simplifies the designers' task considerably.

The heavyweights

The smallest version of the uprated Shuttle would be an unmanned vehicle in which the Orbiter was replaced by a cylindrical cargo container some 24 m long and 7 m wide. It could take a payload of 65 tonnes. The main SSME engines would be housed in a separate pod while the reusable SRBs would be standard ones. The cost per flight of such a launcher is estimated to be about one-third that of today's Shuttle. By substituting liquid-fuel boosters for the SRBs, the payload could be increased still more – to 90 tonnes –

and the launch cost slashed by a third again.

Slightly larger than this version is the one which uses four SRBs rather than the usual two. It could carry cargoes of 90 to 135 tonnes.

Really heavy loads, however, will require an entirely new breed of lifter. Such vehicles are no longer derived from the basic Shuttle. They will be crewless craft but will no longer have the ability to fly back to Earth. They will be rockets whose stages are recovered at the end of their flight, either as they splash down into the sea or softly set down on dry land. The greatest attraction of these squat, dumpy heavyweights is their ability to place up to 300 tonnes of cargo into space at a time, at a cost of somewhere in the region of $50 per kilogram – compared with $900 per kilogram for the present Shuttle.

However, for the time being it appears that any improvement in the Shuttle's payload capacity will be obtained by simply up-rating the three SSMEs and modifying the SRBs.

This chapter describes the Shuttle in detail; first as an integrated whole, then its various parts.

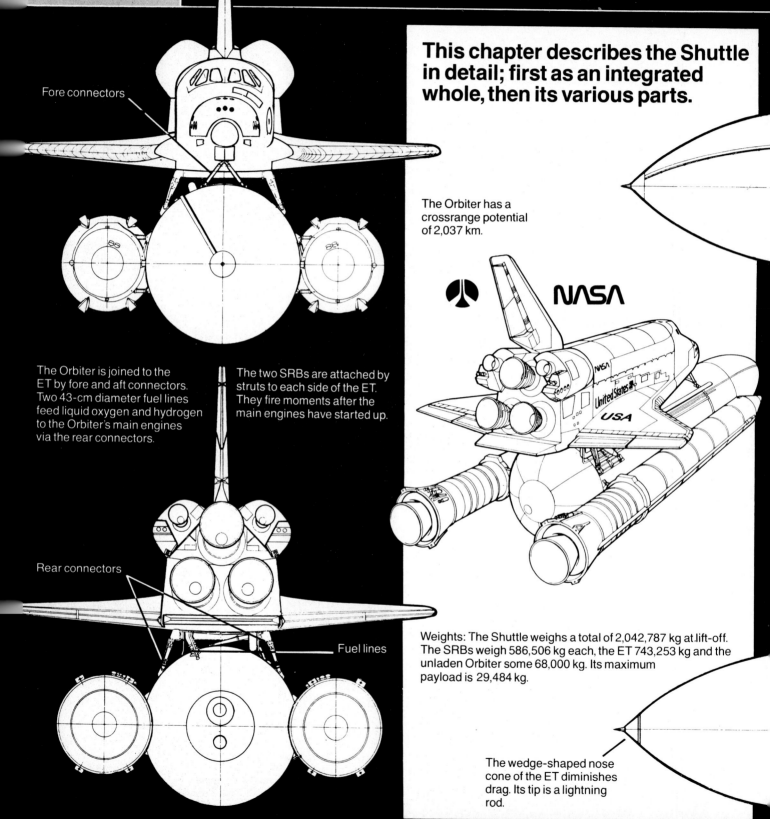

Fore connectors

The Orbiter has a crossrange potential of 2,037 km.

NASA

The Orbiter is joined to the ET by fore and aft connectors. Two 43-cm diameter fuel lines feed liquid oxygen and hydrogen to the Orbiter's main engines via the rear connectors.

The two SRBs are attached by struts to each side of the ET. They fire moments after the main engines have started up.

Rear connectors

Fuel lines

Weights: The Shuttle weighs a total of 2,042,787 kg at lift-off. The SRBs weigh 586,506 kg each, the ET 743,253 kg and the unladen Orbiter some 68,000 kg. Its maximum payload is 29,484 kg.

The wedge-shaped nose cone of the ET diminishes drag. Its tip is a lightning rod.

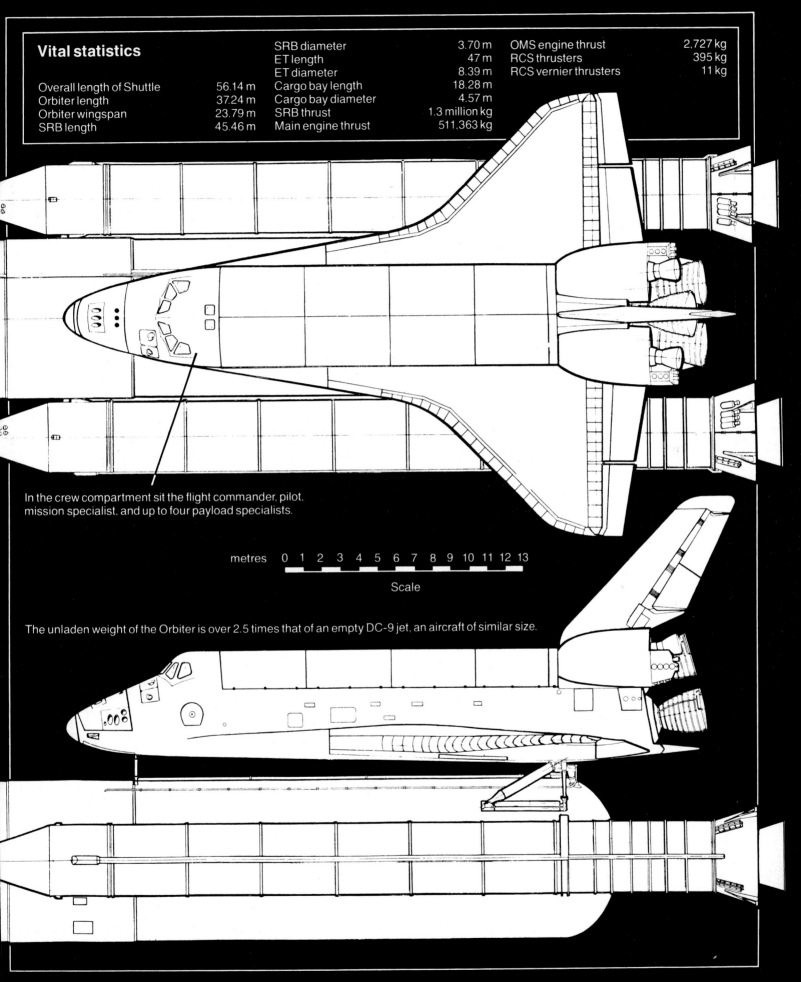

Vital statistics

	SRB diameter	3.70 m	OMS engine thrust	2,727 kg
	ET length	47 m	RCS thrusters	395 kg
	ET diameter	8.39 m	RCS vernier thrusters	11 kg
Overall length of Shuttle	56.14 m	Cargo bay length	18.28 m	
Orbiter length	37.24 m	Cargo bay diameter	4.57 m	
Orbiter wingspan	23.79 m	SRB thrust	1.3 million kg	
SRB length	45.46 m	Main engine thrust	511,363 kg	

In the crew compartment sit the flight commander, pilot, mission specialist, and up to four payload specialists.

metres   0 1 2 3 4 5 6 7 8 9 10 11 12 13

Scale

The unladen weight of the Orbiter is over 2.5 times that of an empty DC-9 jet, an aircraft of similar size.

Solid Rocket Booster (SRB)

Thrust (millions of kg)

1.2
0.8
0.4

30 60 90

Burn time (seconds)

Two thrust vector control units can swivel the nozzle up to 7.1° for flight control during launch.

SRB-ET connector

Nose fairing

Solid propellant

Instrument housing

Main chutes (3)

Nozzle

Drogue chute

The SRBs burn a solid propellant that is a mixture of powdered aluminium perchlorate, aluminium, iron oxide and a rubbery material that holds the mixture together. They consume 503,627 kg of fuel in a burn lasting 123 seconds.

External Tank (ET)

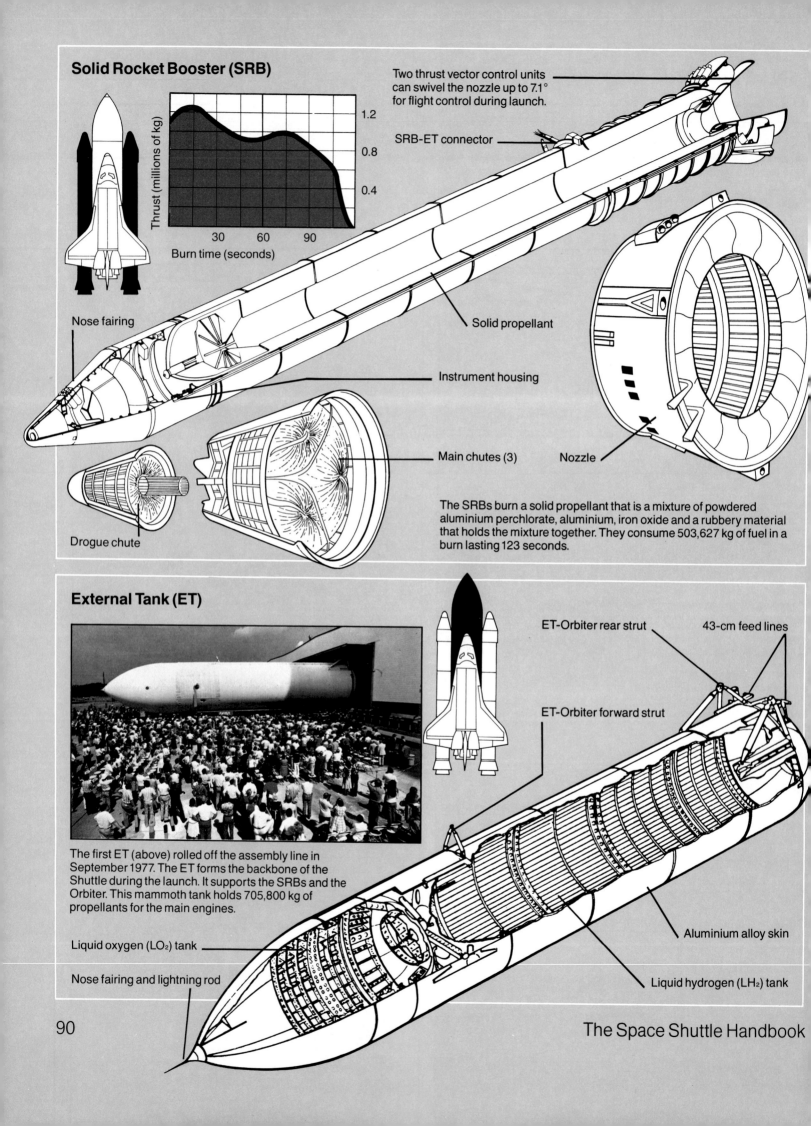

The first ET (above) rolled off the assembly line in September 1977. The ET forms the backbone of the Shuttle during the launch. It supports the SRBs and the Orbiter. This mammoth tank holds 705,800 kg of propellants for the main engines.

ET-Orbiter rear strut

43-cm feed lines

ET-Orbiter forward strut

Aluminium alloy skin

Liquid oxygen (LO₂) tank

Nose fairing and lightning rod

Liquid hydrogen (LH₂) tank

The Space Shuttle Handbook

Space Shuttle Main Engine (SSME)

LH₂ main line

LO₂ main line

LH₂ tank

LO₂ tank

Engines

1
2
3

The three SSMEs are fed by 30-cm pipes that draw fuel off the main line from the ET. Each SSME burns 508 kg of fuel a second at the extremely high pressure of over 200 atmospheres. The engines swivel to give control during the launch.

Engines

30-cm feed pipes

30-cm feed pipes

LO₂ main line

43-cm main line

LH₂ main line

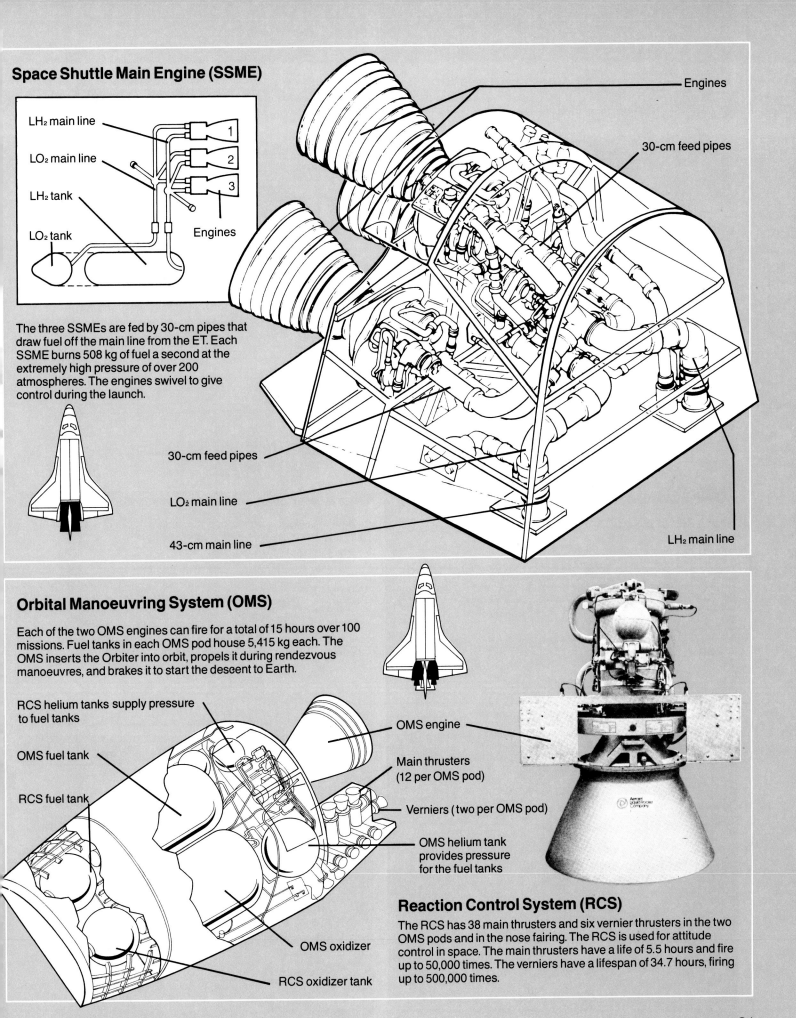

Orbital Manoeuvring System (OMS)

Each of the two OMS engines can fire for a total of 15 hours over 100 missions. Fuel tanks in each OMS pod house 5,415 kg each. The OMS inserts the Orbiter into orbit, propels it during rendezvous manoeuvres, and brakes it to start the descent to Earth.

RCS helium tanks supply pressure to fuel tanks

OMS fuel tank

RCS fuel tank

OMS oxidizer

RCS oxidizer tank

OMS engine

Main thrusters (12 per OMS pod)

Verniers (two per OMS pod)

OMS helium tank provides pressure for the fuel tanks

Reaction Control System (RCS)

The RCS has 38 main thrusters and six vernier thrusters in the two OMS pods and in the nose fairing. The RCS is used for attitude control in space. The main thrusters have a life of 5.5 hours and fire up to 50,000 times. The verniers have a lifespan of 34.7 hours, firing up to 500,000 times.

Plans and Blueprints

Orbiter thermal insulation

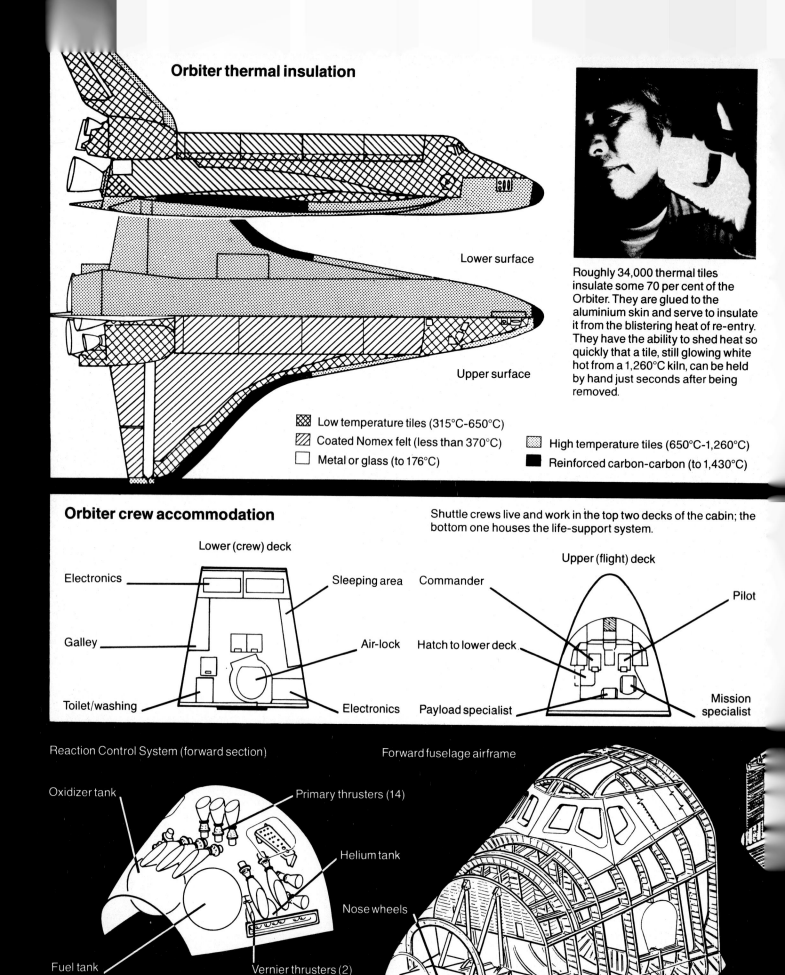

Lower surface

Upper surface

Roughly 34,000 thermal tiles insulate some 70 per cent of the Orbiter. They are glued to the aluminium skin and serve to insulate it from the blistering heat of re-entry. They have the ability to shed heat so quickly that a tile, still glowing white hot from a 1,260°C kiln, can be held by hand just seconds after being removed.

⬚ Low temperature tiles (315°C-650°C)

⬚ Coated Nomex felt (less than 370°C)

⬚ Metal or glass (to 176°C)

⬚ High temperature tiles (650°C-1,260°C)

⬛ Reinforced carbon-carbon (to 1,430°C)

Orbiter crew accommodation

Shuttle crews live and work in the top two decks of the cabin; the bottom one houses the life-support system.

Lower (crew) deck

Electronics

Sleeping area

Galley

Air-lock

Toilet/washing

Electronics

Upper (flight) deck

Commander

Pilot

Hatch to lower deck

Payload specialist

Mission specialist

Reaction Control System (forward section)

Oxidizer tank

Primary thrusters (14)

Helium tank

Nose wheels

Fuel tank

Vernier thrusters (2)

Forward fuselage airframe

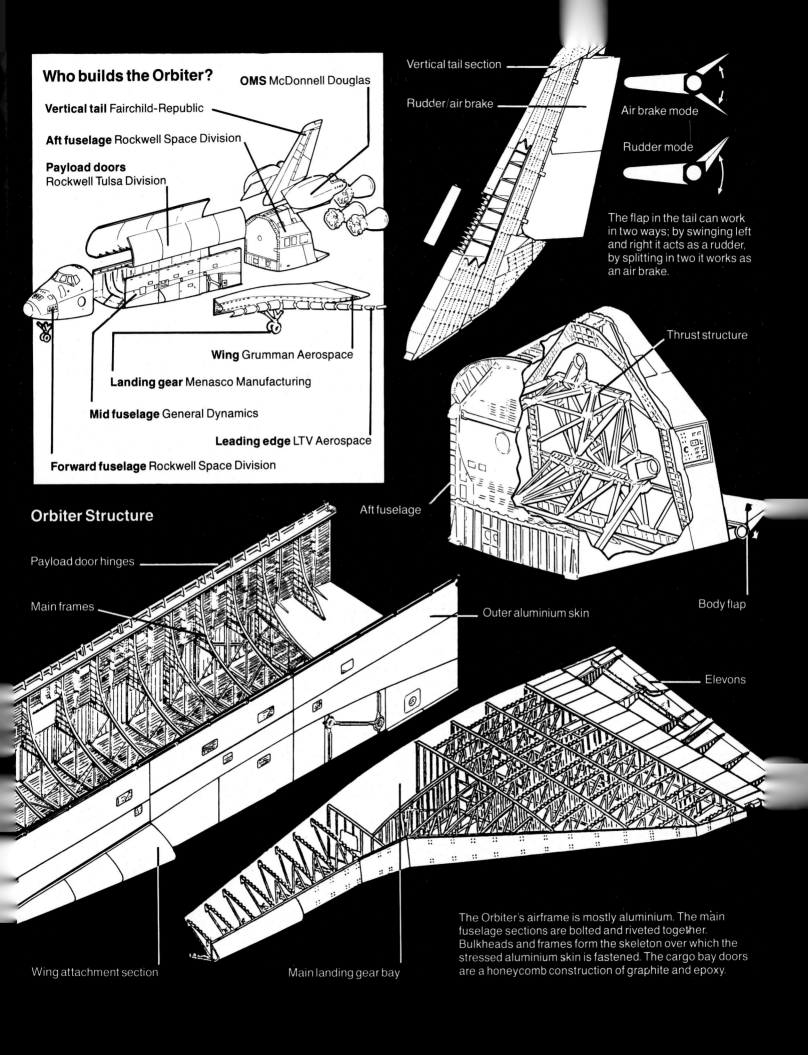

Who builds the Orbiter?

OMS McDonnell Douglas

Vertical tail Fairchild-Republic

Aft fuselage Rockwell Space Division

Payload doors
Rockwell Tulsa Division

Wing Grumman Aerospace

Landing gear Menasco Manufacturing

Mid fuselage General Dynamics

Leading edge LTV Aerospace

Forward fuselage Rockwell Space Division

Vertical tail section

Rudder/air brake

Air brake mode

Rudder mode

The flap in the tail can work in two ways; by swinging left and right it acts as a rudder, by splitting in two it works as an air brake.

Thrust structure

Aft fuselage

Body flap

Orbiter Structure

Payload door hinges

Main frames

Outer aluminium skin

Elevons

Wing attachment section

Main landing gear bay

The Orbiter's airframe is mostly aluminium. The main fuselage sections are bolted and riveted together. Bulkheads and frames form the skeleton over which the stressed aluminium skin is fastened. The cargo bay doors are a honeycomb construction of graphite and epoxy.

Ground turnaround

From touchdown to take-off, the Orbiter can be serviced and re-loaded in the astonishingly short time of 160 hours.

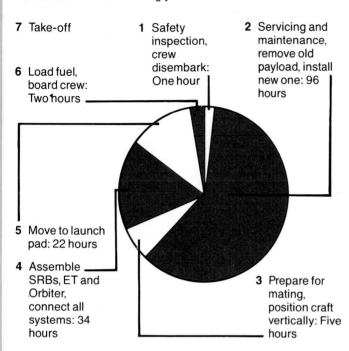

7 Take-off

1 Safety inspection, crew disembark: One hour

2 Servicing and maintenance, remove old payload, install new one: 96 hours

6 Load fuel, board crew: Two hours

5 Move to launch pad: 22 hours

4 Assemble SRBs, ET and Orbiter, connect all systems: 34 hours

3 Prepare for mating, position craft vertically: Five hours

Test flights

The first missions will serve to test the Orbiter's flight performance, its ability to handle payloads and to fly back and land safely. The general outline of some early missions follows.

Mission 1

Orbit inclination: 38°
Altitude: 300 km
Duration: 53 hours
Crew: Two

Testing of SRBs, ET, launch and re-entry procedure, flight control, payload bay doors, single pallet load.

Crossrange 800 km, landing by CSS at Edwards Air Force Base (lake bed).

Mission 3

Orbit inclination: 38°
Altitude: 240 km
Duration: Seven days
Crew: Two

Practice loading and unloading payloads, cargo of sample experiments.

Crossrange 960 km, AUTO approach, CSS landing at Edwards Base (lake bed).

Mission 5

Orbit inclination: 50°
Altitude: 288 km
Duration: Seven days
Crew: Four

Rendezvous testing, manipulating heavy payloads.

Crossrange 1,280 km, AUTO approach, CSS landing at Kennedy Space Center runway.

Spacelab turnaround

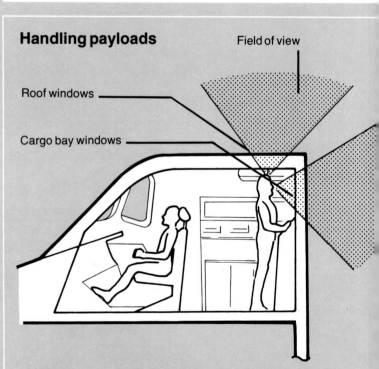

11 Mate Shuttle

12 Move to launch pad

10 Load into Orbiter

13 Launch

9 Check systems

8 Reassemble Spacelab

7 Assemble pallet and rack clusters

Handling payloads

Field of view

Roof windows

Cargo bay windows

Payloads are deployed and retrieved by remote control by an operator working at a station on the rear flight deck.

The robot arm can swing through an arc to reach almost every part of the Orbiter save for the rear fuselage, tail and rear wings. It can deploy a 14,500 kg load in just nine minutes from the bay to a position eight metres above the craft. The overall length of the arm is 15.2 m.

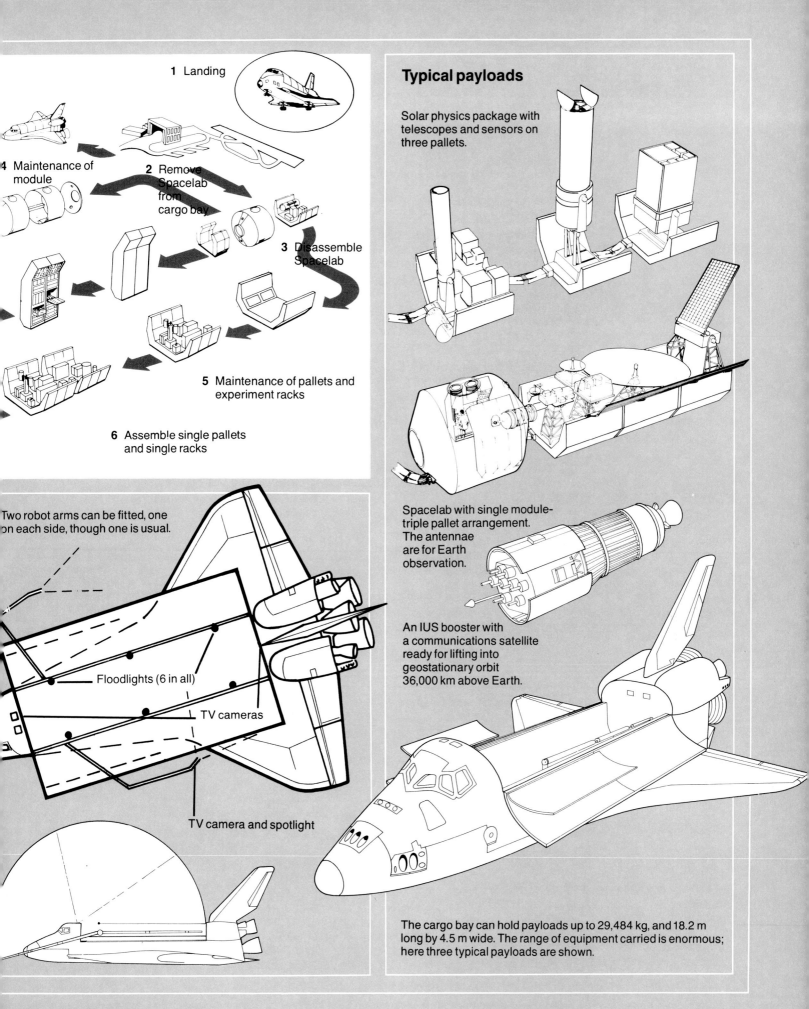

1 Landing

4 Maintenance of module

2 Remove Spacelab from cargo bay

3 Disassemble Spacelab

5 Maintenance of pallets and experiment racks

6 Assemble single pallets and single racks

Typical payloads

Solar physics package with telescopes and sensors on three pallets.

Spacelab with single module-triple pallet arrangement. The antennae are for Earth observation.

An IUS booster with a communications satellite ready for lifting into geostationary orbit 36,000 km above Earth.

Two robot arms can be fitted, one on each side, though one is usual.

Floodlights (6 in all)

TV cameras

TV camera and spotlight

The cargo bay can hold payloads up to 29,484 kg, and 18.2 m long by 4.5 m wide. The range of equipment carried is enormous; here three typical payloads are shown.

INDEX

AFTERWORD

By the turn of the century, the ultimate 'get away from it all' holiday could well be a cruise in space. But it will definitely be a holiday for the rich. The costs of journeying into space are on quite another scale compared with other forms of travel. Whereas the price of flying across the Atlantic is just 90 cents a kilogramme, and a luxury cruise is at the most $11 per kilogramme, a ride into space works out to at least $66 per kilogramme.

Even allowing for the reduced costs of the Shuttle, a ticket would still come to at least $22,500 per trip. However, two decades of technical development, and of inflation, could well bring this figure into the realm of the average first-class tourist.

There is no reason to doubt that there are enough people who are prepared for such an adventure. Several publicity gimmicks to interest the public in space technology drew an overwhelming response. Applications poured in from all over the world in answer to spoof offers of ticket bookings to the Moon.

A journey in an old-fashioned Apollo mooncraft would have demanded exacting physical fitness; but the Shuttle is a different proposition. In fact, Shuttle project managers are convinced that healthy people of any age could make the journey. In principle, there is no reason why package tours into space could not become a reality within our lifetime.